**American Book Company**

*The Standards Experts*

# PASSING THE INDIANA END-OF-COURSE ASSESSMENT IN ALGEBRA I

ERICA DAY

COLLEEN PINTOZZI

AMERICAN BOOK COMPANY

P. O. BOX 2638

WOODSTOCK, GEORGIA 30188-1383

TOLL FREE 1 (888) 264-5877     TOLL FREE FAX 1 (866) 827-3240

WEB SITE: www.americanbookcompany.com

# Acknowledgements

In preparing this book, we would like to acknowledge Samuel Rodriguez and Skylar Saveland for their contributions in writing and editing and Mary Stoddard and Eric Field for their contributions developing graphics for this book. We would also like to thank our many students whose needs and questions inspired us to write this text.

# Contents

Contents

# Contents

# Preface

***Passing the Indiana End-of-Course Assessment in Algebra I*** will help you review and learn important concepts and skills related to algebra 1. When you have finished your review of all of the material your teacher assigns, take the practice tests to evaluate your understanding of the material presented in this book. **The materials in this book are based on the standards in mathematics published by the Indiana Department of Education. The complete list of standards is located on the next page and at the beginning of the Answer Key. Each question in the Practice Tests is referenced to the standard, as is the beginning of each chapter.**

This book contains several sections. These sections are as follows: 1) A Diagnostic Test; 2) Chapters that teach the concepts and skills for the IN Algebra I course; 3) Two Practice Tests. Answers to the tests and exercises are in a separate manual.

# ABOUT THE AUTHORS

**Erica Day** has a Bachelor of Science Degree in Mathematics and working on a Master of Science Degree in Mathematics. She graduated with high honors from Kennesaw State University in Kennesaw, Georgia. She has also tutored all levels of mathematics, ranging from high school algebra and geometry to university-level statistics, calculus, and linear algebra. She is currently writing and editing mathematics books for American Book Company, where she has coauthored numerous books, such as *Passing the Georgia Algebra I End of Course, Passing the Georgia High School Graduation Test in Mathematics, Passing the Arizona AIMS in Mathematics*, and *Passing the New Jersey HSPA in Mathematics*, to help students pass graduation and end of course exams.

**Colleen Pintozzi** has taught mathematics at the middle school, junior high, senior high, and adult level for 22 years. She holds a B.S. degree from Wright State University in Dayton, Ohio and has done graduate work at Wright State University, Duke University, and the University of North Carolina at Chapel Hill. She is the author of many mathematics books including such best-sellers as *Basics Made Easy: Mathematics Review, Passing the New Alabama Graduation Exam in Mathematics, Passing the Louisiana LEAP 21 GEE, Passing the Indiana ISTEP+ GQE in Mathematics, Passing the Minnesota Basic Standards Test in Mathematics*, and *Passing the Nevada High School Proficiency Exam in Mathematics*.

# End-of-Course Assessment Algebra I Reference Sheet

### Pythagorean Theorem

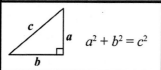

$a^2 + b^2 = c^2$

### Distance Formula

$d = \sqrt{(x_2 - x_1)^2 + (y_2 - y_1)^2}$

$d$ = distance between points 1 and 2

### Midpoint Formula

$M = \left( \frac{x_1 + x_2}{2} , \frac{y_1 + y_2}{2} \right)$

$M$ = point halfway between points 1 and 2

### Standard Form of a Linear Equation

$Ax + By = C$

(where $A$ and $B$ are not both zero)

### Standard Form of a Quadratic Equation

$ax^2 + bx + c = 0$

(where $a \neq 0$)

### Quadratic Formula

$x = \frac{-b \pm \sqrt{b^2 - 4ac}}{2a}$

(where $ax^2 + bx + c = 0$ and $a \neq 0$)

### Equation of a Line

**Slope-Intercept Form:** $y = mx + b$
where $m$ = slope and $b$ = $y$-intercept

**Point-Slope Form:** $y - y_1 = m(x - x_1)$

### Simple Interest Formula

$I = prt$

where $I$ = interest

$p$ = principal

$r$ = rate

$t$ = time

### Slope of a Line

Let $(x^1, y^1)$ and $(x^2, y^2)$ be two points in a plane

slope = $\frac{\text{change in } y}{\text{change in } x} = \frac{y^2 - y^1}{x^2 - x^1}$

(where $x_2 \neq x_1$)

| Shape | | Formulas for Area (*A*) and Circumference (*C*) | |
|---|---|---|---|
| **Triangle** | | $A = \frac{1}{2}bh = \frac{1}{2} \times$ base $\times$ height | |
| **Trapezoid** | | $A = \frac{1}{2}(b_1 + b_2)h = \frac{1}{2} \times$ sum of bases $\times$ height | |
| **Parallelogram** | | $A = bh =$ base $\times$ height | |
| **Circle** | | $A = \pi r^2 = \pi \times$ square of radius <br> $C = 2\pi r = 2 \times \pi \times$ radius | $\pi \approx 3.14$ <br> or $\pi \approx \frac{22}{7}$ |
| **Figure** | | Formulas for Volume (*V*) and Surface Area (*SA*) | |
| **Cube** | | $SA = 6s^2 = 6 \times$ length of side squared | |
| **Cylinder (total)** | | $SA = 2\pi rh + 2\pi r^2$ <br> $SA = 2 \times \pi \times$ radius $\times$ height $+ 2 \times \pi \times$ radius squared | $\pi \approx 3.14$ <br> or <br> $\pi \approx \frac{22}{7}$ |
| **Sphere** | | $SA = 4\pi r^2 = 4 \times \pi \times$ radius squared <br> $V = \frac{4}{3}\pi r^3 = \frac{4}{3} \times \pi \times$ radius cubed | |
| **Cone** | | $V = \frac{1}{3}\pi r^2 h = \frac{1}{3} \times \pi \times$ radius squared $\times$ height | |
| **Pyramid** | | $V = \frac{1}{3}Bh = \frac{1}{3} \times$ area of base $\times$ height | |
| **Prism** | | $V = Bh =$ area of base $\times$ height | |

# Diagnostic Test

## Session 1

1. Simplify: $\dfrac{x^2 - 4}{x^2 + x - 6}$

   A $\dfrac{x-2}{x+3}$

   B $\dfrac{x+2}{x+3}$

   C $\dfrac{x+2}{x-2}$

   D $\dfrac{x-2}{x-3}$

   A1.7.1

2. A kindergarten teacher needs to buy 10 markers for each student to color with. She has 19 kids in her class. If markers come in packs of 6, what is the minimum number of packs she must buy?

   A 32

   B 31

   C 3

   D 2

   A1.2.6

3. Simplify the following expression.

   $\dfrac{x^2 - 4}{x^2 + 5x + 6}$

   A 1

   B $\dfrac{x+2}{x+3}$

   C $\dfrac{x-2}{x+2}$

   D $\dfrac{x-2}{x+3}$

   A1.7.1

4. Is the following graph a function?

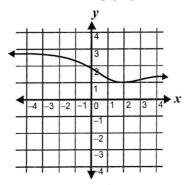

   A Yes, because it passes the vertical line test.

   B Yes, because it passes the horizontal line test.

   C No, because it fails the vertical line test.

   D No, because it fails the horizontal line test.

   A1.3.3

5. Solve for $b$. $4b + a = 13$

   A $13 - 4a$

   B $13 - 4b$

   C $\dfrac{13 - a}{4}$

   D $\dfrac{13 - b}{a}$

   A1.2.2

6. Solve the proportion.

   $\dfrac{2x - 1}{3} = \dfrac{x + 4}{6}$

   A $x = 6$

   B $x = 3$

   C $x = 2$

   D $x = 1$

   A1.7.2

1

7. What is the domain of the following function?

$$y = x^3$$

**A** $0 \le x < \infty$
**B** $3 \le x < \infty$
**C** $-\infty < x < \infty$
**D** $0 < x < \infty$

A1.3.4

8. What is the range of the following function?

$$y = x^2$$

**A** $0 \le y < \infty$
**B** $0 \le x < \infty$
**C** $-\infty < y < \infty$
**D** $-\infty < x < \infty$

A1.3.4

9. Where do these two equations intersect?

$$3x + y = 3$$

$$2y - x = -8$$

**A** $(0, 1)$
**B** $(0, 3)$
**C** $(2, 3)$
**D** $(2, -3)$

A1.5.5

10. Find the solution for the pair of equations.

$$2x + 4y = 40$$

$$x = 3y$$

**A** $(6, 4)$
**B** $(12, 4)$
**C** $(4, 6)$
**D** $(4, 12)$

A1.5.3

11. Convert 37 kilometers per hour to meters per second.

A1.1.5

12. Where does this pair of equations intersect?

$$y = -x$$

$$3x + 2y = -4$$

**A** $(-4, -4)$
**B** $(4, -4)$
**C** $(-4, 4)$
**D** $(4, 4)$

A1.5.5

13. Solve the equation $(x + 6)^2 = 169$

**A** $x = 5, 13$
**B** $x = -17, 3$
**C** $x = -19, 7$
**D** $x = -7, 7$

A1.8.3

14. Solve the equation $\sqrt{4x - 1} = 2x$

**A** $x = \frac{1}{2}$
**B** $x = -\frac{1}{2}, \frac{1}{2}$
**C** $x = -\frac{1}{4}, 0$
**D** $x = -1, 1$

A1.8.8

15. Solve $3x^2 + x - 2 = 0$ by factoring.

**A** $x = -2, 3$
**B** $x = 1, 3$
**C** $x = -\frac{2}{3}, 0$
**D** $x = -1, \frac{2}{3}$

A1.8.2

16. Solve the combined inequality

$$\frac{-x - 6}{3} \ge \frac{-4x}{3} > \frac{-27 - x}{3}.$$

**A** $-1 < x \le 5$
**B** $-9 < x \le -2$
**C** $3 < x < 7$
**D** $2 \le x < 9$

A1.2.5

17. Simplify the ratio $\dfrac{2x^2 + 2x - 60}{6x^2 + 48x + 72}$.

   **A** $\dfrac{x + 6}{x + 2}$

   **B** $\dfrac{x - 5}{3x + 6}$

   **C** $\dfrac{2x - 10}{3x + 6}$

   **D** $\dfrac{x - 5}{x - 6}$

A1.7.1

18. Solve the algebraic proportion.
$$\frac{x - 6}{2} = \frac{3x + 2}{4}$$

   **A** $x = -4$

   **B** $x = -14$

   **C** $x = -2$

   **D** $x = -8$

A1.7.2

19. Divide $6xyz + 2xz + 7yz + 2z$ by $2z$.

   **A** $3xz + 2x + 7yz + 1$

   **B** $\dfrac{xy}{2} + x + \dfrac{7y}{2} + \dfrac{z}{2}$

   **C** $3xy + x + \dfrac{7y}{2} + 1$

   **D** $\dfrac{6xy}{z} + \dfrac{x}{2} + 3y + \dfrac{2}{z}$

A1.6.5

20. Simplify $3\left(-x^2 + 4x + 1\right) - 2\left(3x^2 + 2x - 2\right)$.

   **A** $-3x^2 - 16x + 1$

   **B** $-6x^2 + 12x - 4$

   **C** $-9x^2 + 8x + 7$

   **D** $-12x^2 + 16x - 3$

A1.6.1

21. Graph the following inequality.

   $-9x + 3y < 3$

A1.4.6

22. Multiply the polynomials $(6x + 2)(x - 1)$.

   **A** $6x^2 - 4x - 2$

   **B** $6x^2 - 8x + 3$

   **C** $2x^2 + 3x - 2$

   **D** $12x^2 + 8x + 4$

A1.6.4

23. Which of these graphs represents the inequality $y \geq 2x + 1$?

**A**

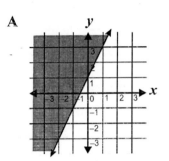

**B**

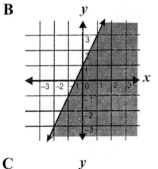

**C**

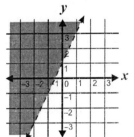

**D**

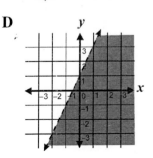

A1.4.6

24. A single bag of fertilizer can cover about 500 ft$^2$ of grass. Derik owns a farm that is 4,350,000 ft$^2$. Write an inequality that shows how many bags of fertilizer Derik needs for all his land.

A1.2.6

25. A high school football player is practicing his field goal kicks. The equation below represents the height of the ball at a specific time.

$s = -9t^2 + 45t$, where $t$ is the number of seconds gone by and $s$ is the height of the ball in feet.

How long does it take for the ball to hit the ground?

A1.8.7

26. Graph the equations. Where do the two lines intersect?

$-4x + 2y = -6$ and $-3x + 3y = 0$

A1.5.1

27. Simplify: $\sqrt{75}$

**A** $5\sqrt{3}$
**B** $3\sqrt{5}$
**C** $3\sqrt{25}$
**D** $5\sqrt{15}$

A1.1.2

28. Evaluate: $9^{\frac{3}{2}}$

**A** 3
**B** 81
**C** 27
**D** 122.5

A1.1.4

# Session 2

1. What is the equation $3x - 6y = 5$ in slope-intercept form?

   A   $3x - 6y = 5$

   B   $3x - 6y - 5 = 0$

   C   $y = \frac{1}{2}x - \frac{5}{6}$

   D   $3x = 6y + 5$

   A1.4.3

2. Which equation runs through the points $(3, -1)$ and $(6, 0)$?

   A   $y = \frac{1}{3}x - 2$

   B   $y = x - 4$

   C   $y = 2x$

   D   $y = x$

   A1.4.4

3. What is $\sqrt{720}$ in simplest form?

   A   $3\sqrt{80}$

   B   $10\sqrt{72}$

   C   $10\sqrt{3}$

   D   $12\sqrt{5}$

   A1.1.2

4. Which number is the largest? $6^2$, $3^4$, $37$, or $\sqrt{1225}$

   A   $6^2$

   B   $3^4$

   C   $37$

   D   $\sqrt{1225}$

   A1.1.1

5. Simplify $16^{\frac{3}{2}}$

   A   $16$

   B   $4$

   C   $4096$

   D   $64$

   A1.1.4

6. Which graph represents the equation $-12x + 6y = -6$?

   A
   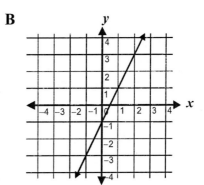

   B
   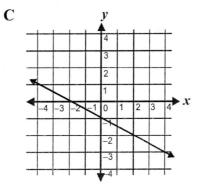

   C

   D
   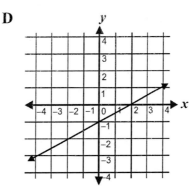

   A1.4.1

7. What term can be factored out of $24x^3y^2 - 18x^2y^2 + 12xy^6$?

   **A** $6x^2y^2$

   **B** $6xy^2$

   **C** $x^3y$

   **D** $xy^2$

                        A1.6.6

8. Factor $2x^2 - 2x - 112$

   **A** $2(x+7)(x-8)$

   **B** $(2x-7)(2x-8)$

   **C** $(2x-5)(x+10)$

   **D** $3(x+3)(x-2)$

                        A1.6.7

9. Solve the equation $(x+6)^2 = 81$ by completing the square.

   **A** $x = -5, 15$

   **B** $x = -3, 9$

   **C** $x = -3, 3$

   **D** $x = -15, 3$

                        A1.8.3

10. Solve $x^2 + 2x - 35 = 0$ by factoring.

   **A** $x = -3, 5$

   **B** $x = -7, 5$

   **C** $x = -4, 6$

   **D** $x = -4, 9$

                        A1.8.2

11. Solve the equation $\sqrt{4x + 21} = x$.

   **A** $x = -7, 3$

   **B** $x = -3, 7$

   **C** $x = -7, -3$

   **D** $x = 3, 7$

                        A1.8.8

12. Find the common solution for the pair of equations $2x + 6y = 17$ and $2x + 4y = 13$.

   **A** $(-4.5, 5.5)$

   **B** $(1, 2.5)$

   **C** $(-2, 3.5)$

   **D** $(2.5, 2)$

                        A1.5.4

13. Which graph represents the equation $2y = 3x + 4$?

**A**

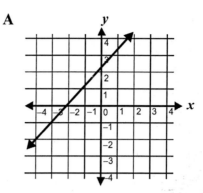

**B**

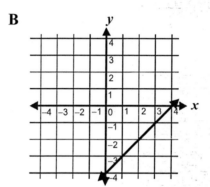

**C**

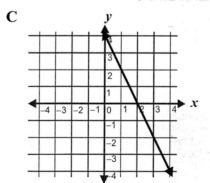

**D**

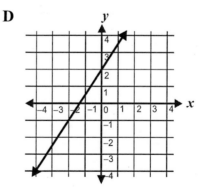

                        A1.4.1

14. Which is the graph of the inequality $x + y \geq 4$?

**A**

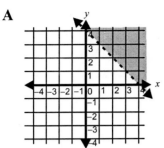

**B**

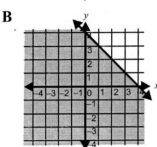

**C**

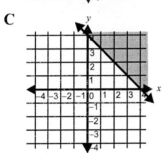

**D**

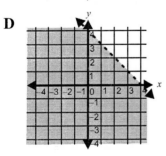

A1.4.6

15. What is the $y$-intercept of the equation $6x + y = 18$?

**A** $(0, 18)$
**B** $(0, 3)$
**C** $(18, 0)$
**D** $(3, 0)$

A1.4.2

16. Given the points $(-2, 5)$ and $(3, -5)$, find the equation of the line.

**A** $y = -3x + 1$
**B** $y = -2x + 1$
**C** $y = x - 8$
**D** $y = -3x + 4$

A1.4.4

17. Solve $6x^2 - 24x - 270$ using the quadratic formula.

**A** $x = -5, 9$
**B** $x = 6, 15$
**C** $x = -7, 5$
**D** $x = -9, 3$

A1.8.6

18. Solve the inequality $11x - 6 \geq -3x - 20$.

**A** $x \geq -2$
**B** $x \geq -1$
**C** $x \leq 7$
**D** $x \leq 3$

A1.2.4

19. Which numbers in the set $\{-7, -2, 0, 3, 4, 6\}$ make the inequality $2x + 4 \leq 8$ true?

**A** $\{-7, -2, 0, 3, 4, 6\}$
**B** $\{-7, 0, 3, 4, 6\}$
**C** $\{-7, -2, 3, 4\}$
**D** $\{-7, -2, 0\}$

A1.2.3

20. For the function $y = x^3 + x^2$, what is the range of the function?

**A** $-\infty < y < \infty$
**B** $0 \leq y < \infty$
**C** $-\infty < y \leq 0$
**D** $-\infty < x < \infty$

A1.3.4

21. Is $y^2 = x$ a function? Explain your reasoning.

   **A** Yes, because it passes the horizontal line test.

   **B** Yes, because it passes the vertical line test.

   **C** No, because it fails horizontal line test.

   **D** No, because it fails the vertical line test.

                A1.3.3

22. Solve the pair of linear equations $x = 2y + 3$ and $2x + 5y = 42$.

                A1.5.3

23. The graph below shows a correlation between the number of students there are and the number of sandwiches eaten.

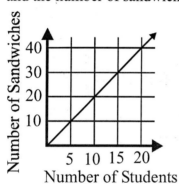

What is the number of sandwiches eaten if there were 77 students?

                A1.4.5

24. Kate's SUV gets 21 miles per gallon of gasoline. If she has to drive somewhere that is 273 miles away, write an inequality that shows how much gasoline Kate must use in order to reach her destination.

                A1.2.6

25. A restaurant buys $600 worth of hamburgers, hamburger buns, and condiments. If the restaurant is selling hamburgers for $1.75, how many hamburgers must they sell to begin making profit?

                A1.2.6

26. Solve $x^2 + 6x - 7 = 0$ by completing the square. Explain each step.

                A1.8.4

# Evaluation Chart for the Diagnostic Mathematics Test

**Directions:** On the following chart, circle the question numbers that you answered incorrectly. Then turn to the appropriate topics (listed by chapters), read the explanations, and complete the exercises. Review the other chapters as needed. Finally, complete the *Passing the Indiana End-of-Course Assessment in Algebra I* Practice Tests to further review.

| | | Questions Session 1 | Questions Session 2 | Pages |
|---|---|---|---|---|
| Chapter 1: | Exponents | | | 10–20 |
| Chapter 2: | Roots | 27, 28 | 3, 4, 5 | 21–30 |
| Chapter 3: | Solving Multi-Step Linear Equations and Inequalities | 5, 16 | 18, 19 | 31–45 |
| Chapter 4: | Algebra Word Problems | 2, 11, 24 | 24, 25 | 46–60 |
| Chapter 5: | Polynomials | 19, 20, 22 | | 61–78 |
| Chapter 6: | Factoring | 1, 3, 6, 17, 18 | 7, 8 | 79–94 |
| Chapter 7: | Solving Quadratic Equations | 13, 14, 15, 25 | 9, 10, 11, 17, 26 | 95–106 |
| Chapter 8: | Graphing and Writing Equations and Inequalities | 21, 23 | 1, 2, 6, 13, 14, 15, 16 | 107–123 |
| Chapter 9: | Applications of Graphs | | 23 | 124–140 |
| Chapter 10: | Pairs of Linear Equations and Inequalities | 9, 10, 12, 26 | 12, 22 | 141–155 |
| Chapter 11: | Relations and Functions | 4, 7, 8 | 20, 21 | 156–170 |
| Chapter 12: | Mathematical Reasoning | | | 171–185 |

# Chapter 1
# Exponents

This chapter covers the following IN Algebra I standards:

| Standard 1: | Operations with Real Numbers | A1.1.4 |
| Standard 9: | Mathematical Reasoning and Problem Solving | A1.9.3 |

## 1.1 Understanding Exponents

Sometimes it is necessary to multiply a number by itself one or more times. For example, a math problem may need to multiply $3 \times 3$ or $5 \times 5 \times 5 \times 5$. In these situations, mathematicians have come up with a shorter way of writing out this kind of multiplication. Instead of writing $3 \times 3$, you can write $3^2$, or instead of writing $5 \times 5 \times 5 \times 5$, $5^4$ means the same thing. The first number is the **base**. The small, raised number is called the **exponent** or **power**. The exponent tells how many times the base should be multiplied by itself.

**Example 1:**   $6^3$ ← **exponent (or power)**
              ← **base**   This means multiply by 6 three times: $6 \times 6 \times 6$

**Example 2:**   $4^1 = 4$   $10^1 = 10$   $25^1 = 25$   $4^0 = 1$   $10^0 = 1$   $25^0 = 1$

All rational numbers can have exponents.

**Examples:**   $\left(\dfrac{1}{4}\right)^3 = \dfrac{1}{4} \times \dfrac{1}{4} \times \dfrac{1}{4} = \dfrac{1}{64}$   $0.2^3 = 0.2 \times 0.2 \times 0.2 = 0.008$

**Rewrite the following problems using <u>exponents</u>.**

**Example 3:** $2 \times 2 \times 2 = 2^3$

1. $7 \times 7 \times 7 \times 7$
2. $10 \times 10$
3. $12 \times 12 \times 12$
4. $4 \times 4 \times 4 \times 4$
5. $9 \times 9 \times 9$
6. $25 \times 25$

**Use your calculator to determine what product each number with an exponent represents.**

**Example 4:** $2^3 = 2 \times 2 \times 2 = 8$

7. $8^3$
8. $12^2$
9. $20^1$
10. $5^4$
11. $15^0$
12. $16^2$
13. $10^2$
14. $3^5$

**Express each of the following numbers as a base with an exponent.**

**Example 5:** $4 = 2 \times 2 = 2^2$

15. 9
16. 16
17. 27
18. 36
19. 8
20. 32
21. 1000
22. 125

## 1.2   Multiplying Exponents with the Same Base

To multiply two expressions with the same base, add the exponents together and keep the base the same.

**Example 6:**   $2^3 \times 2^5 = (2 \times 2 \times 2)(2 \times 2 \times 2 \times 2 \times 2) = 2^8$

**Example 7:**   $3a^2 \times 2a^3 = 6a^{2+3} = 6a^5$
Notice that only the "$a$" is raised to a power and not the 3 or the 2.

**Simplify each of the expressions below.**

1. $2^3 \times 2^5$

2. $x^5 \times x^3$

3. $2a^3 \times 3a^3$

4. $4^5 \times 4^3$

5. $2x^3 \times x^5$

6. $4b^3 \times 2b^4$

7. $10^5 \times 10^4$

8. $5^2 \times 5^4$

9. $3^3 \times 3^2$

10. $4x \times x^2$

11. $a^2 \times 3a^4$

12. $2^3 \times 2^4$

## 1.3   Multiplying Fractional Exponents with the Same Base

To multiply two expressions with the same base, add the exponents together and keep the base the same. Numbers with **fractional exponents** follow the same rules as numbers with whole numbers as the exponent.

**Example 8:**   $(4)^{\frac{1}{2}} \times (4)^{\frac{3}{2}} = 4^{\frac{1}{2}+\frac{3}{2}} = 4^{\frac{4}{2}} = 4^2$

**Example 9:**   $2x^{\frac{3}{4}} \times 5x^{\frac{1}{5}} = 10x^{\frac{3}{4}+\frac{1}{5}} = 10x^{\frac{15}{20}+\frac{4}{20}} = 10x^{\frac{19}{20}}$

Notice that only the "$x$" is raised to a power and not the 2 or the 5.

**Simplify each of the expressions below.**

1. $y^{\frac{1}{3}} \times y^{\frac{2}{3}}$

2. $(2)^{\frac{5}{9}} \times (2)^{\frac{1}{3}}$

3. $(10)^{\frac{4}{7}} \times (10)^{\frac{5}{7}}$

4. $b^{\frac{1}{2}} \times b^{\frac{1}{2}}$

5. $x^{\frac{1}{6}} \times x^{\frac{2}{3}}$

6. $(6)^{\frac{1}{12}} \times (6)^{\frac{3}{4}}$

7. $(3)^{\frac{5}{12}} \times (3)^{\frac{7}{2}}$

8. $(12)^{\frac{13}{25}} \times (12)^{\frac{1}{5}}$

9. $2a^{\frac{7}{9}} \times a^{\frac{4}{9}}$

10. $(8)^{\frac{2}{5}} \times (8)^{\frac{1}{10}}$

11. $4a^{\frac{6}{7}} \times 4a^{\frac{4}{5}}$

12. $3(x)^{\frac{4}{13}} \times 5(x)^{\frac{1}{2}}$

## 1.4   Multiplying Exponents Raised to an Exponent

If a power is raised to another power, multiply the exponents together and keep the base the same.

**Example 10:**   $(2^3)^2 = (2^3)(2^3) = (2 \times 2 \times 2)(2 \times 2 \times 2) = 2^{3 \times 2} = 2^6$

**Example 11:**   $(y^4)^3 = y^{4 \times 3} = y^{12}$

**Simplify each of the expressions below.**

1. $(5^3)^2$    4. $(3^4)^2$    7. $(3^3)^2$    10. $(a^4)^2$

2. $(x^5)^2$    5. $(3^2)^4$    8. $(9^2)^2$    11. $(5^4)^5$

3. $(6^2)^5$    6. $(y^2)^3$    9. $(7^2)^2$    12. $(x^3)^2$

## 1.5   Expressions Raised to a Power

If an expression is raised to a power, do not raise each term to the power, but rather consider the expression as a whole and raise it to the power.

**Example 12:**   $(2+3)^2 = (2+3)(2+3) = (5)(5) = 25$

**Example 13:**   $(2-5)^3 = (2-5)(2-5)(2-5) = (-3)(-3)(-3) = -27$

**Simplify each of the expressions below.**

1. $(2+6)^2$    6. $(2+3)^3$    11. $(5-9+2)^3$

2. $(1-2)^2$    7. $(4-6)^2$    12. $(2+1+4)^2$

3. $(3+2)^2$    8. $(2+1)^4$    13. $(-7+-7+6)^3$

4. $(4+5)^2$    9. $(25+15)^0$    14. $(2+2-5+4)^2$

5. $(5+1)^3$    10. $(4+1+7)^2$    15. $(20+45+40)^0$

## 1.6  Fractions Raised to a Power

A fraction can also be raised to a power.

**Example 14:**  $\left(\dfrac{3}{4}\right)^3 = \dfrac{3^3}{4^3} = \dfrac{27}{64}$

**Simplify the following fractions.**

1. $\left(\dfrac{2}{3}\right)^2$

4. $\left(\dfrac{1}{4}\right)^2$

7. $\left(\dfrac{1}{2}\right)^3$

10. $\left(\dfrac{3}{10}\right)^2$

2. $\left(\dfrac{7}{8}\right)^3$

5. $\left(\dfrac{2}{3}\right)^3$

8. $\left(\dfrac{5}{7}\right)^2$

11. $\left(\dfrac{4}{5}\right)^2$

3. $\left(\dfrac{1}{2}\right)^2$

6. $\left(\dfrac{3}{4}\right)^2$

9. $\left(\dfrac{2}{3}\right)^4$

12. $\left(\dfrac{1}{10}\right)^4$

## 1.7  More Multiplying Exponents

If a product in parentheses is raised to a power, then each factor is raised to the power when parentheses are eliminated.

**Example 15:**  $(2 \times 4)^2 = 2^2 \times 4^2 = 4 \times 16 = 64$

**Example 16:**  $(3a)^3 = 3^3 \times a^3 = 27a^3$

**Example 17:**  $(7b^5)^2 = 7^2 b^{10} = 49b^{10}$

**Simplify each of the following.**

1. $(2^3)^2$

2. $(7a^5)^2$

3. $(6b^2)^2$

4. $(3^2)^2$

5. $(3 \times 5)^2$

6. $(3x^4)^2$

7. $(6y^7)^2$

8. $(11w^3)^2$

9. $(3^3)^2$

10. $(3 \times 3)^2$

11. $(2a)^4$

12. $(2^2)^3$

13. $(3 \times 2)^3$

14. $(5^3)^2$

15. $(4r^7)^3$

16. $(2m^3)^2$

17. $(6 \times 4)^2$

18. $(9a^5)^2$

19. $(5x^4)^2$

20. $(9^2)^2$

21. $4 \times 4^3$

22. $(3a)^2$

23. $(2 \times 3)^3$

24. $(5p^4)^3$

25. $(4y^4)^2$

26. $(2b^3)^4$

27. $(5a^2)^2$

28. $(8a^3)^2$

29. $(2 \times 6)^2$

30. $(7^2)^2$

## 1.8   Negative Exponents

Expressions can also have negative exponents. Negative exponents do not indicate negative numbers. They indicate **reciprocals**. The **reciprocal** of a number is 1 divided by that number. For example, the reciprocal of 2 is $\frac{1}{2}$. (A number multiplied by its reciprocal is equal to 1.) If the negative exponent is in the bottom of a fraction (denominator), the reciprocal will put the expression on the top of the fraction (numerator) without a negative sign.

**Example 18:**   $2^{-3} = \frac{1}{2^3} = \frac{1}{8}$

**Example 19:**   $3a^{-5} = 3 \times \frac{1}{a^5} = \frac{3}{a^5}$  Notice that the 3 is not raised to the $-5$ power, only the $a$.

**Example 20:**   $\frac{6}{5x^{-2}} = \frac{6x^2}{5}$   The 5 is not raised to the $-2$ power, only the $x$.

**Rewrite using only positive exponents.**

1. $5m^{-6}$

2. $\frac{5x^{-4}}{7}$

3. $14z^{-8}$

4. $\frac{1}{5s^{-4}}$

5. $14h^{-5}$

6. $\frac{h^{-3}}{5}$

7. $\frac{2y^{-3}}{4}$

8. $x^{-4}$

9. $-2y^{-2}$

10. $5y^{-5}$

11. $\frac{x^{-3}}{5}$

12. $10z^{-7}$

13. $7x^{-3}$

14. $r^{-2}$

15. $\frac{m^{-4}}{6}$

## 1.9   Multiplying with Negative Exponents

Multiplying with negative exponents follows the same rules as multiplying with positive exponents.

**Example 21:**   $6^2 \cdot 6^{-3} = 6^{2+(-3)} = 6^{-1} = \frac{1}{6}$

**Example 22:**   $(5a \times 2)^{-3} = (10a)^{-3} = \frac{1}{(10a)^3} = \frac{1}{1000a^3}$

**Example 23:**   $(7a^2)^{-3} = 7^{-3}a^{-6} = \frac{1}{7^3a^6}$

**Simplify the following. Answers should <u>not</u> have any negative exponents.**

1. $5^{-2} \cdot 5^5$

2. $(6^3 \cdot 6^{-2})^{-2}$

3. $10^{-4} \cdot 10^2$

4. $11^{-5} \cdot 11^7$

5. $4^7 \cdot 4^{-10}$

6. $20^8 \cdot 20^{-6}$

7. $5^{-8} \cdot 5^4$

8. $(2^{-2} \cdot 2^3)^{-4}$

9. $7^{-2} \cdot 7^{-1}$

10. $(3x^4)^{-3}$

11. $12^{-10} \cdot 12^8$

12. $(10^8 \cdot 10^{-10})^2$

13. $3^{-2} \cdot 2^{-2}$

14. $(8x^5)^{-4}$

15. $(6b^3)^{-6}$

16. $(9y)^{-2}$

## 1.10   Dividing with Exponents

Exponents that have the same base can also be divided.

**Example 24:**   $\dfrac{3^5}{3^3}$   This problem means $3^5 \div 3^3$. Let us look at 2 ways to solve this problem.

**Solution 1:**   $\dfrac{3^5}{3^3} = \dfrac{3 \cdot 3 \cdot 3 \cdot 3 \cdot 3}{3 \cdot 3 \cdot 3} = 3 \cdot 3 = 9$   First, rewrite the fraction with the exponents in expanded form, and then multiply.

**Solution 2:**   $\dfrac{3^5}{3^3} = 3^{5-3} = 3^2 = 9$   A quick way to simplify this same problem is to subtract the exponents. **When dividing exponents with the same base, subtract the exponents.**

**Example 25:**   $\dfrac{(4x)^{-3}}{2x^4}$

**Step 1:**   $(4x)^{-3} = \dfrac{1}{(4x)^3} = \dfrac{1}{4^3 x^3}$   Remove the parentheses from the top of the fraction.

**Step 2:**   $\dfrac{1}{4^3 x^3 \cdot 2x^4} = \dfrac{1}{128x^7}$   The bottom of the fraction remains the same, so put the two together and simplify.

**Simplify the problems below. You may be able to cancel. Be sure to follow order of operations. Remove parentheses before canceling.**

1. $\dfrac{5^5}{5^3}$

2. $\dfrac{x^2}{x^3}$

3. $\dfrac{(10^2)^4}{10^5}$

4. $\dfrac{3^5}{3^2}$

5. $\dfrac{8^{10}}{8^8}$

6. $\dfrac{5^2}{5}$

7. $\dfrac{(7^2)^3}{7^5}$

8. $\dfrac{(x^3)^4}{x^6}$

9. $\dfrac{4^3}{4^2}$

10. $\dfrac{2}{(2^2)^2}$

11. $\dfrac{(3x)^{-2}}{9x^5}$

12. $\dfrac{(11^4)^4}{(11^7)^2}$

13. $\dfrac{x^3}{(x^2)^3}$

14. $\dfrac{2^2}{2^7}$

15. $\dfrac{6^2}{6}$

16. $\dfrac{9^{11}}{9^9}$

17. $\dfrac{(15)^5}{15^6}$

18. $\dfrac{(x^3)^{-2}}{(x^2)^5}$

19. $\dfrac{12^{-4}}{12^{-2}}$

20. $\dfrac{6^{12}}{6^9}$

21. $\dfrac{8^8}{8^{10}}$

22. $\dfrac{3(x^{-3})^{-2}}{3x^7}$

23. $\dfrac{7^3}{7^5}$

24. $\dfrac{10^3}{10^{-1}}$

## 1.11   Order of Operations

In long math problems with $+$, $-$, $\times$, $\div$, (), and exponents in them, you have to know what to do first. Without following the same rules, you could get different answers. If you will memorize the silly sentence, Please Excuse My Dear Aunt Sally, you can memorize the order you must follow.

**P**lease     "**P**" stands for parentheses. You must get rid of parentheses first.
Examples: $3(1+4) = 3(5) = 15$
$6(10-6) = 6(4) = 24$

**E**xcuse     "**E**" stands for exponents. You must eliminate exponents next.
Example: $4^2 = 4 \times 4 = 16$

**M**y **D**ear     "**M**" stands for multiply. "**D**" stands for divide. Start on the left of the equation and perform all multiplications and divisions in the order in which they appear.

**A**unt **S**ally     "**A**" stands for add. "**S**" stands for subtract. Start on the left and perform all additions and subtractions in the order they appear.

---

**Example 26:** $12 \div 2(6-3) + 3^2 - 1$

Please     Eliminate **parentheses**. $6-3 = 3$ so now we have     $12 \div 2 \times 3 + 3^2 - 1$

Excuse     Eliminate **exponents**. $3^2 = 9$ so now we have     $12 \div 2 \times 3 + 9 - 1$

My Dear     **Multiply** and **divide** next in order from left to right.     $12 \div 2 = 6$ then $6 \times 3 = 18$

Aunt Sally     Last, we **add** and **subtract** in order from left to right.     $18 + 9 - 1 = 26$

---

**Simplify the following problems.**

1. $6 + 9 \times 2 - 4$

2. $3(4+2) - 6^2$

3. $3(6-3) - 2^3$

4. $49 \div 7 - 3 \times 3$

5. $10 \times 4 - (7-2)$

6. $2 \times 3 \div 6 \times 4$

7. $4^3 \div 8(4+2)$

8. $7 + 8(14-6) \div 4$

9. $(2+8-12) \times 4$

10. $4(8-13) \times 4$

11. $8 + 4^2 \times 2 - 6$

12. $3^2(4+6) + 3$

13. $(12-6) + 27 \div 3^2$

14. $82^0 - 1 + 4 \div 2^2$

15. $1 - (2-3) + 8$

16. $-4\{18 - (4 + 2 \times 6)\}$

17. $18 \div (6+3) - 12$

18. $10^2 + 3^3 - 2 \times 3$

19. $4^2 + (7+2) \div 3$

20. $7 \times 4 - 9 \div 3$

When a problem has a fraction bar, simplify the top of the fraction (numerator) and the bottom of the fraction (denominator) separately using the rules for order of operations. You treat the top and bottom as if they were separate problems. Then reduce the fraction to lowest terms.

**Example 27:** $\dfrac{2(4-3)-6}{5^2+3(2+1)}$

| Please | Eliminate **parentheses**. $(4-3)=1$ and $(2+1)=3$ | $\dfrac{2\times 1-6}{5^2+3\times 3}$ |
|---|---|---|
| Excuse | Eliminate **exponents**. $5^2=25$ | $\dfrac{2\times 1-6}{25+3\times 3}$ |
| My Dear | **Multiply** and **divide** in the numerator and denominator separately. $3\times 3=9$ and $2\times 1=2$ | $\dfrac{2-6}{25+9}$ |
| Aunt Sally | **Add** and **subtract** in the numerator and denominator separately. $2-6=-4$ and $25+9=34$ | $\dfrac{-4}{34}$ |

Now reduce the fraction to lowest terms. $\dfrac{-4}{34}=\dfrac{-2}{17}$

**Simplify the following problems.**

1. $\dfrac{2^2+4}{5+3(8+1)}$

2. $\dfrac{8^2-(4+11)}{4^2-3^2}$

3. $\dfrac{5-2(4-3)}{2(1-8)}$

4. $\dfrac{10+(2-4)}{4(2+6)-2^2}$

5. $\dfrac{3^3-8(1+2)}{-10-(3+8)}$

6. $\dfrac{(9-3)+3^2}{-5-2(4+1)}$

7. $\dfrac{16-3(10-6)}{(13+15)-5^2}$

8. $\dfrac{(2-5)-11}{12-2(3+1)}$

9. $\dfrac{7+(8-16)}{6^2-5^2}$

10. $\dfrac{16-(12-3)}{8(2+3)-5}$

11. $\dfrac{-3(9-7)}{7+9-2^3}$

12. $\dfrac{4-(2+7)}{13+(6-9)}$

13. $\dfrac{5(3-8)-2^2}{7-3(6+1)}$

14. $\dfrac{3(3-8)+5}{8^2-(5+9)}$

15. $\dfrac{6^2-4(7+3)}{8+(9-3)}$

# Chapter 1 Review

**Simplify the following expressions. Simplify the answers. Make all exponents positive.**

1. $5^2 \times 5^3$

2. $\left(4^4\right)^5$

3. $\left(4y^3\right)^3$

4. $6x^{-3}$

5. $\left(3a^2\right)^{-2}$

6. $\left(b^3\right)^{-4}$

7. $\dfrac{4^6}{4^4}$

8. $\left(\dfrac{3}{5}\right)^2$

9. $x^3 \cdot x^{-7}$

10. $(2x)^{-4}$

11. $3^3 \times 3^2$

12. $\left(2^4\right)^2$

13. $5^7 \times 5^{-4}$

14. $\dfrac{\left(3a^2\right)^3}{a^3}$

15. $\left(4^2\right)^{-2}$

16. $\left(5^{-9} \times 5^7\right)^{-2}$

17. $\dfrac{\left(2^3\right)^2}{2^4}$

18. $\dfrac{y^{-2}}{3y^4}$

19. $(6x)^{-3}$

20. $\left(4d^5\right)^{-3}$

21. $(5)^{\frac{3}{4}} \times (5)^{\frac{1}{3}}$

22. $(6)^{\frac{1}{12}} \times (6)^{\frac{1}{4}}$

**Write using exponents.**

23. $3 \times 3 \times 3 \times 3$

24. $6 \times 6 \times 6 \times 6 \times 6 \times 6$

25. $11 \times 11 \times 11$

26. $2 \times 2 \times 2 \times 2 \times 2 \times 2 \times 2 \times 2$

**Simplify the following problems using the correct order of operations.**

27. $10 \div (-1 - 4) + 2$

28. $5 + (2)(4 - 1) \div 3$

29. $5 - 5^2 + (2 - 4)$

30. $(8 - 10) \times (5 + 3) - 10$

31. $\dfrac{10 + 5^2 - 3}{2^2 + 2(5 - 3)}$

32. $1 - (9 - 1) \div 2$

33. $\dfrac{5(3 - 6) + 3^2}{4(2 + 1) - 6}$

34. $-4(6 + 4) \div (-2) + 1$

35. $12 \div (7 - 4) - 2$

36. $1 + 4^2 \div (3 + 1)$

# Chapter 1 Test

1. Simplify the expression shown below:

$$\frac{8x^4}{2x^2}$$

**A** $2x^4$

**B** $4x^2$

**C** $\dfrac{1}{4x^2}$

**D** $\dfrac{4x^2}{x}$

2. Simplify the following:

$$5 \cdot x^4 \cdot y^5 \cdot z^{-3}$$

**A** $\dfrac{5x^4y^5}{z^3}$

**B** $(5xyz)^6$

**C** $\dfrac{625x^4y^5}{z^3}$

**D** $x^{20}y^{25}z^{-15}$

3. What is the solution to $2(5-2)^2 - 15 \div 5$?

**A** $-\frac{3}{5}$

**B** $\frac{3}{5}$

**C** 15

**D** $4\frac{1}{5}$

4. Simplify: $(6)^{\frac{5}{6}} \times (6)^{\frac{1}{3}}$

**A** $(6)^{\frac{7}{6}}$

**B** $(6)^{\frac{5}{18}}$

**C** $(6)^{\frac{6}{9}}$

**D** $(36)^{\frac{7}{6}}$

5. Simplify: $3^2 + 4 \times 18 \div 9$

**A** 4

**B** 14

**C** 17

**D** 26

6. $4 \times 18 - 9 \div 3^2 =$

**A** 4

**B** 7

**C** 68

**D** 71

7. Simplify: $\dfrac{(4x)^{-3}}{6x^4}$

**A** $\dfrac{2}{3x^{12}}$

**B** $\dfrac{2}{3x^7}$

**C** $\dfrac{1}{384x^7}$

**D** $-\dfrac{32}{3x}$

8. Simplify: $\left(6x^4\right)^{-2}$

**A** $\dfrac{1}{36x^8}$

**B** $-36x^8$

**C** $-12x^{-8}$

**D** $6x^{-8}$

9. $x^2 \cdot x^4 =$

**A** $x^8$

**B** $8x$

**C** $x^6$

**D** $6x$

10. Write using exponents: $4a \times 4a \times 4a$

  **A** $4a^3$
  **B** $64a^3$
  **C** $3(4a)$
  **D** $(4+a)^3$

11. Simplify: $(3^4)^2$

  **A** $3^8$
  **B** $3^6$
  **C** $12^2$
  **D** $7^2$

12. $(5y)^{-4} =$

  **A** $\dfrac{1}{(5y)^4}$

  **B** $\dfrac{1}{5y^4}$

  **C** $-20 - 4y$

  **D** $\dfrac{5}{y^4}$

13. $3^3 x^2 \cdot 4x^5 =$

  **A** $108x^{10}$
  **B** $108x^5$
  **C** $108x^7$
  **D** $12x^{10}$

14. $3^2 \cdot 3^{-3} =$

  **A** $9^{-6}$

  **B** $\dfrac{1}{3}$

  **C** $\dfrac{1}{9}$

  **D** $3$

15. $8y^{-2} =$

  **A** $\dfrac{1}{8y^2}$

  **B** $6y$

  **C** $-16y$

  **D** $\dfrac{8}{y^2}$

16. $(4+2)^3 =$

  **A** $18$
  **B** $72$
  **C** $216$
  **D** $54$

# Chapter 2
# Roots

This chapter covers the following IN Algebra I standards:

| Standard 1: Operations with Real Numbers | A1.1.1 |
| --- | --- |
| | A1.1.2 |
| | A1.1.4 |

## 2.1   Square Root

Just as working with exponents is related to multiplication, finding square roots is related to division. In fact, the sign for finding the square root of a number looks similar to a division sign. The best way to learn about square roots is to look at examples.

**Examples:**   This is a square root problem: $\sqrt{64}$
It is asking, "What is the square root of 64?"
It means, "What number multiplied by itself equals 64?"
The answer is 8. $8 \times 8 = 64$.

**Find the square roots of the following numbers.**

1. $\sqrt{49}$     4. $\sqrt{16}$     7. $\sqrt{100}$     10. $\sqrt{36}$     13. $\sqrt{64}$

2. $\sqrt{81}$     5. $\sqrt{121}$     8. $\sqrt{289}$     11. $\sqrt{4}$     14. $\sqrt{9}$

3. $\sqrt{25}$     6. $\sqrt{625}$     9. $\sqrt{196}$     12. $\sqrt{900}$     15. $\sqrt{144}$

## 2.2   Simplifying Square Roots Using Factors

Square roots can sometimes be simplified even if the number under the square root is not a perfect square. One of the rules of roots is that if $a$ and $b$ are two positive real numbers, then it is always true that $\sqrt{a \cdot b} = \sqrt{a} \cdot \sqrt{b}$. You can use this rule to simplify square roots.

**Example 1:**     $\sqrt{100} = \sqrt{4 \cdot 25} = \sqrt{4} \cdot \sqrt{25} = 2 \cdot 5 = 10$

**Example 2:**     $\sqrt{200} = \sqrt{100 \cdot 2} = 10\sqrt{2} \longleftarrow$ Means 10 multiplied by the square root of 2

**Example 3:**     $\sqrt{160} = \sqrt{10 \cdot 16} = 4\sqrt{10}$

**Simplify.**

1. $\sqrt{98}$     3. $\sqrt{50}$     5. $\sqrt{8}$     7. $\sqrt{48}$     9. $\sqrt{54}$     11. $\sqrt{72}$     13. $\sqrt{90}$     15. $\sqrt{18}$

2. $\sqrt{600}$     4. $\sqrt{27}$     6. $\sqrt{63}$     8. $\sqrt{75}$     10. $\sqrt{40}$     12. $\sqrt{80}$     14. $\sqrt{175}$     16. $\sqrt{20}$

## 2.3    Adding, Subtracting, and Simplifying Square Roots

You can add and subtract terms with square roots only if the number under the square root sign is the same.

**Example 4:**    $2\sqrt{2} + 3\sqrt{2} = 5\sqrt{2}$

**Example 5:**    $12\sqrt{7} - 3\sqrt{7} - 9\sqrt{7}$

Or, look at the following examples where you can simplify the square roots and then add or subtract.

**Example 6:**    $2\sqrt{25} + \sqrt{36}$

**Step 1:**    Simplify. You know that $\sqrt{25} = 5$, and $\sqrt{36} = 6$ so the problem simplifies to $2(5) + 6$

**Step 2:**    Solve: $2(5) + 6 = 10 + 6 = 16$

**Example 7:**    $2\sqrt{72} - 3\sqrt{2}$

**Step 1:**    Simplify what you know. $\sqrt{72} = \sqrt{36 \cdot 2} = 6\sqrt{2}$

**Step 2:**    Substitute $6\sqrt{2}$ for $\sqrt{72}$ simplify.
$2(6)\sqrt{2} - 3\sqrt{2} = 12\sqrt{2} - 3\sqrt{2} = 9\sqrt{2}$

**Simplify the following addition and subtraction problems.**

1. $3\sqrt{5} + 9\sqrt{5}$

2. $3\sqrt{25} + 4\sqrt{16}$

3. $4\sqrt{8} + 2\sqrt{2}$

4. $3\sqrt{32} - 2\sqrt{2}$

5. $\sqrt{25} - \sqrt{49}$

6. $2\sqrt{5} + 4\sqrt{20}$

7. $5\sqrt{8} - 3\sqrt{72}$

8. $\sqrt{27} + 3\sqrt{27}$

9. $3\sqrt{20} - 4\sqrt{45}$

10. $4\sqrt{45} - \sqrt{75}$

11. $2\sqrt{28} + 2\sqrt{7}$

12. $\sqrt{64} + \sqrt{81}$

13. $5\sqrt{54} - 2\sqrt{24}$

14. $\sqrt{32} + 2\sqrt{50}$

15. $2\sqrt{7} + 4\sqrt{63}$

16. $8\sqrt{2} + \sqrt{8}$

17. $2\sqrt{8} - 4\sqrt{32}$

18. $\sqrt{36} + \sqrt{100}$

19. $\sqrt{9} + \sqrt{25}$

20. $\sqrt{64} - \sqrt{36}$

21. $\sqrt{75} + \sqrt{108}$

22. $\sqrt{81} + \sqrt{100}$

23. $\sqrt{192} - \sqrt{75}$

24. $3\sqrt{5} + \sqrt{245}$

## 2.4 Multiplying and Simplifying Square Roots

You can also multiply square roots. To multiply square roots, you just multiply the numbers under the square root sign and then simplify. Look at the examples below.

**Example 8:** $\sqrt{2} \times \sqrt{6}$

**Step 1:** $\sqrt{2} \times \sqrt{6} = \sqrt{2 \times 6} = \sqrt{12}$     Multiply the numbers under the square root sign.

**Step 2:** $\sqrt{12} = \sqrt{4 \times 3} = 2\sqrt{3}$     Simplify

**Example 9:** $3\sqrt{3} \times 5\sqrt{6}$

**Step 1:** $(3 \times 5)\sqrt{3 \times 6} = 15\sqrt{18}$     Multiply the numbers in front of the square root, and multiply the numbers under the square root sign.

**Step 2:** $15\sqrt{18} = 15\sqrt{2 \times 9}$
$15 \times 3\sqrt{2} = 45\sqrt{2}$     Simplify.

**Example 10:** $\sqrt{14} \times \sqrt{42}$     For this more complicated multiplication problem, use the rule of roots that you learned on page 21, $\sqrt{a \cdot b} = \sqrt{a} \cdot \sqrt{b}$.

**Step 1:** $\sqrt{14} = \sqrt{7} \times \sqrt{2}$ and
$\sqrt{42} = \sqrt{2} \times \sqrt{3} \times \sqrt{7}$     Instead of multiplying 14 by 42, divide these numbers into their roots.

$\sqrt{14} \times \sqrt{42} = \sqrt{7} \times \sqrt{2} \times \sqrt{2} \times \sqrt{3} \times \sqrt{7}$

**Step 2:** Since you know that $\sqrt{7} \times \sqrt{7} = 7$ and $\sqrt{2} \times \sqrt{2} = 2$, the problem simplifies to $(7 \times 2)\sqrt{3} = 14\sqrt{3}$

**Simplify the following multiplication problems.**

1. $\sqrt{5} \times \sqrt{7}$

2. $\sqrt{32} \times \sqrt{2}$

3. $\sqrt{10} \times \sqrt{14}$

4. $2\sqrt{3} \times 3\sqrt{6}$

5. $4\sqrt{2} \times 2\sqrt{10}$

6. $\sqrt{5} \times 3\sqrt{15}$

7. $\sqrt{45} \times \sqrt{27}$

8. $5\sqrt{21} \times \sqrt{7}$

9. $\sqrt{42} \times \sqrt{21}$

10. $4\sqrt{3} \times 2\sqrt{12}$

11. $\sqrt{56} \times \sqrt{24}$

12. $\sqrt{11} \times 2\sqrt{33}$

13. $\sqrt{13} \times \sqrt{26}$

14. $2\sqrt{2} \times 5\sqrt{5}$

15. $\sqrt{6} \times \sqrt{12}$

## 2.5   Dividing and Simplifying Square Roots

When dividing a number or a square root by another square root, you cannot leave the square root sign in the denominator (the bottom number) of a fraction. You must simplify the problem so that the square root is not in the denominator. This is also called rationalizing the denominator. Look at the examples below.

**Example 11:**   $\dfrac{\sqrt{2}}{\sqrt{5}}$

**Step 1:**   $\dfrac{\sqrt{2}}{\sqrt{5}} \times \dfrac{\sqrt{5}}{\sqrt{5}}$ ⟵   The fraction $\frac{\sqrt{5}}{\sqrt{5}}$ is equal to 1, and multiplying by 1 does not change the value of a number.

**Step 2:**   $\dfrac{\sqrt{2 \times 5}}{5} = \dfrac{\sqrt{10}}{5}$   Multiply and simplify. Since $\sqrt{5} \times \sqrt{5}$ equals 5, you no longer have a square root in the denominator.

**Example 12:**   $\dfrac{6\sqrt{2}}{2\sqrt{10}}$   In this problem, the numbers outside of the square root will also simplify.

**Step 1:**   $\dfrac{6}{2} = 3$ so you have $\dfrac{3\sqrt{2}}{\sqrt{10}}$

**Step 2:**   $\dfrac{3\sqrt{2}}{\sqrt{10}} \times \dfrac{\sqrt{10}}{\sqrt{10}} = \dfrac{3\sqrt{2 \times 10}}{10} = \dfrac{3\sqrt{20}}{10}$

**Step 3:**   $\dfrac{3\sqrt{20}}{10}$ will further simplify because $\sqrt{20} = 2\sqrt{5}$, so you then have $\dfrac{3 \times 2\sqrt{5}}{10}$ which reduces to $\dfrac{3\sqrt{5}}{5}$.

**Simplify the following division problems.**

1. $\dfrac{9\sqrt{3}}{\sqrt{5}}$

2. $\dfrac{16}{\sqrt{8}}$

3. $\dfrac{24\sqrt{10}}{12\sqrt{3}}$

4. $\dfrac{\sqrt{121}}{\sqrt{6}}$

5. $\dfrac{\sqrt{40}}{\sqrt{90}}$

6. $\dfrac{33\sqrt{15}}{11\sqrt{2}}$

7. $\dfrac{\sqrt{32}}{\sqrt{12}}$

8. $\dfrac{\sqrt{11}}{\sqrt{5}}$

9. $\dfrac{\sqrt{2}}{\sqrt{6}}$

10. $\dfrac{2\sqrt{7}}{\sqrt{14}}$

11. $\dfrac{5\sqrt{2}}{4\sqrt{8}}$

12. $\dfrac{4\sqrt{21}}{7\sqrt{7}}$

13. $\dfrac{9\sqrt{22}}{2\sqrt{2}}$

14. $\dfrac{\sqrt{35}}{2\sqrt{14}}$

15. $\dfrac{\sqrt{40}}{\sqrt{15}}$

16. $\dfrac{\sqrt{3}}{\sqrt{12}}$

## 2.6   Cube Roots

**Cube roots** look like square roots, except that there is a "3" raised in front of the root sign:

Square root of 64: $\sqrt{64}$

Cube root of 64: $\sqrt[3]{64}$

In fact, they function very much like square roots, with one important difference. Recall asking, "What is the square root of 64?" means:

"What number multiplied by itself equals 64?"

Asking "What is the cube root of 64?" means:

"What number multiplied 3 times ('cubed') by itself equals 64?"

The answer is 4. $4 \times 4 \times 4 = 64$.

**Find the cube root of the following numbers.**

**Examples:** $\sqrt[3]{27}$        $3 \times 3 \times 3 = 27$ so $\sqrt[3]{27} = 3$

$\sqrt[3]{1000}$     $10 \times 10 \times 10 = 1000$ so $\sqrt[3]{1000} = 10$

**Find the cube roots of the following numbers.**

1. $\sqrt[3]{1}$

2. $\sqrt[3]{8}$

3. $\sqrt[3]{64}$

4. $\sqrt[3]{125}$

5. $\sqrt[3]{27}$

6. $\sqrt[3]{\frac{64}{27}}$

7. $\sqrt[3]{1000}$

8. $\sqrt[3]{\frac{125}{1000}}$

## 2.7   Rational Exponents

Sometimes exponents can be rational numbers.

**Example 13:**
$$4^{\frac{3}{2}}$$
$3 \longleftarrow$ the exponent of the number under the radical
$2 \longleftarrow$ the root you need to take

$$4^{\frac{3}{2}} = \sqrt[2]{4^3} = \sqrt{4^3} = \sqrt{64} = 8$$

**Note:** To solve these problems you can raise the base to the exponent *before* or *after* you take the root.

**Example 14:**
$$\sqrt{2} = 2^{\frac{1}{2}}$$

$$\sqrt[3]{10} = 10^{\frac{1}{3}}$$

$$\sqrt[3]{3^2} = 3^{\frac{2}{3}}$$

$$\sqrt[5]{4^2} = 4^{\frac{2}{5}}$$

**Example 15:**   Evaluate $\sqrt[10]{7^{20}}$ using rational exponents.

$$\sqrt[10]{7^{20}} = 7^{\frac{20}{10}} = 7^2 = 7 \times 7 = 49$$

**Change the following to an expression using a radical and find its integer equivalent.**

1. $25^{\frac{3}{2}}$

2. $16^{\frac{5}{2}}$

3. $4^{\frac{5}{2}}$

4. $36^{\frac{3}{2}}$

5. $8^{\frac{2}{3}}$

6. $125^{\frac{1}{3}}$

7. $32^{\frac{3}{5}}$

8. $4^{\frac{1}{2}}$

9. $25^{\frac{5}{2}}$

10. $64^{\frac{2}{3}}$

11. $27^{\frac{2}{3}}$

12. $81^{\frac{3}{4}}$

**Change the following expressions to a whole number base with a rational exponent.**

13. $\sqrt{5}$

14. $\sqrt[3]{19}$

15. $\sqrt[4]{2^3}$

16. $\sqrt[5]{9^2}$

17. $\sqrt[3]{20}$

18. $\sqrt[4]{11^3}$

19. $\sqrt{5^3}$

20. $\sqrt[4]{7^3}$

21. $\sqrt[5]{2^3}$

22. $\sqrt{50}$

23. $\sqrt[3]{8^2}$

24. $\sqrt[5]{9^3}$

25. $\sqrt[10]{5^{30}}$

26. $\sqrt{2^6}$

27. $\sqrt[4]{7^5}$

28. $\sqrt[5]{12^2}$

**Find the value of each expression.**

29. $\left(\sqrt{10}\right)^4$

30. $\sqrt[9]{10^{27}}$

31. $\sqrt{1,307,428^2}$

32. $\sqrt{3^4}$

# 2.8  Comparing Real Number Expressions

Expressions such as $2^3$, $\sqrt{49}$, $8.5^{\frac{1}{3}}$, and $10^{-2}$ represent real numbers. Every real number expression can be evaluated or approximated. Every real number expression will be 'greater than'($>$), 'less than' ($<$), or 'equal to' ($=$) any other real number expression.

Approximating irrational numbers like $\sqrt{2}$ with enough precision to compare can be technically challenging. If the two numbers are not too close together, the easiest way to compare expressions is to convert them to a decimal.

**Example 16:**  Which is greater, $2^3$ or $\sqrt{81}$?

    **Step 1:**  Evaluating both expressions. We see that $2^3 = 2 \times 2 \times 2 = 8$ and $\sqrt{81} = 9$.

    **Step 2:**  Compare. Clearly, $8 < 9$. So, $2^3 < \sqrt{81}$.

**Example 17:**  Which is greater, $8.5^{\frac{1}{3}}$ or $10^{-2}$?

    **Step 1:**  Find what $8.5^{\frac{1}{3}}$ equals. What is the cube root of 8.5? What number multiplied by itself three times, $x \times x \times x$ equals 8.5? We know that $2^3 = 8$ and $3^3 = 27$. So, the cube root of 8.5 is somewhere between 2 and 3. Technology shows that this number is approximately equal to 2.04. So, $8.5^{\frac{1}{3}} \approx 2.04$

    **Step 2:**  Find what $10^{-2}$ equals. $10^{-2} = \dfrac{1}{10^2} = \dfrac{1}{10 \times 10} = \dfrac{1}{100} = 0.1$

    **Step 3:**  Compare the real numbers. Which is bigger? 2.04 or 0.1? It's not hard to see that $0.1 < 2.04$. So, $10^{-2} < 8.5^{\frac{1}{3}}$.

**Use the symbols, $>$, $<$, and $=$ to compare the following expressions.**

1. $5^2$ and 30

2. $\sqrt{144}$ and $3^2$

3. $\frac{3}{4}$ and $(0.5)^{10}$

4. $1 + 9$ and $9 + 10^{-2}$

5. $10^2$ and $2^{10}$

6. $(10^2)^{10}$ and $\sqrt{1,000,000}$

7. $0.4729$ and $\frac{17}{32}$

8. 1 and $\sqrt{1}$

9. $1^{100}$ and $100^1$

10. $\frac{9}{2}$ and $8.21 \div 2.1$

11. $\sqrt{2}$ and $2^{-2}$

12. $\frac{1}{10}$ and $\left(\frac{2}{10}\right)^2$

13. $3.1415$ and $\frac{22}{7}$

14. $1^{-100}$ and $(-10)^2$

15. $\frac{1}{4}$ and $(0.5)^2$

16. $0.\overline{99}$ and 1

# Chapter 2 Review

**Simplify the following square root expressions.**

1. $\sqrt{50}$

2. $\sqrt{44}$

3. $\sqrt{12}$

4. $\sqrt{18}$

5. $\sqrt{8}$

6. $\sqrt{48}$

7. $\sqrt{75}$

8. $\sqrt{200}$

9. $\sqrt{32}$

10. $\sqrt{20}$

11. $\sqrt{63}$

12. $\sqrt{80}$

**Simplify the following square root problems.**

13. $5\sqrt{27} + 7\sqrt{3}$

14. $\sqrt{40} - \sqrt{10}$

15. $\sqrt{64} + \sqrt{81}$

16. $8\sqrt{50} - 3\sqrt{32}$

17. $14\sqrt{5} + 8\sqrt{80}$

18. $\sqrt{63} \times \sqrt{28}$

19. $\dfrac{\sqrt{56}}{\sqrt{35}}$

20. $\sqrt{8} \times \sqrt{50}$

21. $\dfrac{\sqrt{20}}{\sqrt{45}}$

22. $5\sqrt{40} \times 3\sqrt{20}$

23. $2\sqrt{48} - \sqrt{12}$

24. $\dfrac{2\sqrt{5}}{\sqrt{30}}$

25. $\dfrac{3\sqrt{22}}{2\sqrt{3}}$

26. $\sqrt{72} \times 3\sqrt{27}$

27. $4\sqrt{5} + 8\sqrt{45}$

**Rewrite each of the following with rational (fractional) exponents.**

28. $\sqrt[3]{2^4}$

29. $\sqrt[5]{1^2}$

30. $\sqrt[4]{8^3}$

31. $\sqrt[3]{6^2}$

32. $\sqrt[6]{2^{12}}$

33. $\sqrt[5]{9^3}$

**Find the cube roots of the following numbers.**

34. $\sqrt[3]{512}$

35. $\sqrt[3]{\dfrac{125}{8}}$

**Find the value of each expression.**

36. $9^{\frac{3}{2}}$

37. $100^{\frac{3}{2}}$

38. $4^{\frac{1}{2}}$

39. $\left(\sqrt[3]{3}\right)^6$

**Use the symbols, $>$, $<$, and $=$ to compare the following expressions.**

40. $3.25^2$ and $\sqrt{100}$

41. $10^3$ and $5^6$

42. $0.95^{50}$ and $0.99^2$

43. $\left(\frac{3}{2}\right)^2$ and $\left(\frac{2}{3}\right)^3$

# Chapter 2 Test

1. Simplify: $\sqrt{135}$

   A $3\sqrt{15}$

   B $\sqrt{72}$

   C $9\sqrt{15}$

   D $\sqrt{9} \times \sqrt{15}$

2. Express $\dfrac{\sqrt{20}}{\sqrt{35}}$ in simplest form.

   A $\dfrac{2\sqrt{7}}{7}$

   B $\dfrac{2}{\sqrt{7}}$

   C $\dfrac{2\sqrt{5}}{\sqrt{7}}$

   D $\dfrac{4}{7}$

3. Simplify: $\dfrac{3\sqrt{12}}{2\sqrt{3}}$

   A $3$

   B $\dfrac{3\sqrt{4}}{2}$

   C $\dfrac{6}{\sqrt{6}}$

   D $3\sqrt{3}$

4. Simplify: $\sqrt{44} \cdot 2\sqrt{33}$

   A $2\sqrt{77}$

   B $44\sqrt{3}$

   C $22\sqrt{7}$

   D $22\sqrt{12}$

5. Rewrite with rational exponents:

   $\sqrt[5]{3^2}$

   A $3^{10}$

   B $5^{\frac{2}{3}}$

   C $3^{\frac{5}{2}}$

   D $3^{\frac{2}{5}}$

6. Simplify: $\sqrt{45} \times \sqrt{27}$

   A $3\sqrt{15}$

   B $\sqrt{72}$

   C $9\sqrt{15}$

   D $\sqrt{9} \times \sqrt{15}$

7. Simplify the following by rationalizing the denominator.

   $\dfrac{\sqrt{3}}{\sqrt{15}}$

   A $\dfrac{\sqrt{5}}{5}$

   B $\dfrac{1}{\sqrt{5}}$

   C $\dfrac{\sqrt{45}}{15}$

   D $\dfrac{3\sqrt{5}}{15}$

8. Simplify: $\sqrt[3]{40}$

   A $8\sqrt[3]{5}$

   B $\sqrt[3]{8} \times \sqrt[3]{5}$

   C $2\sqrt[3]{5}$

   D You cannot take the cube root of 40.

9. Which is equivalent to $\sqrt[9]{2^3}$?

   **A** $\frac{8}{9}$

   **B** $2^{\frac{1}{3}}$

   **C** 1.2

   **D** 27

10. Which is equivalent to $5^{\frac{2}{9}}$?

   **A** $\sqrt[9]{25}$

   **B** $\frac{25}{9}$

   **C** $\sqrt[3]{125}$

   **D** $\left(\sqrt{5}\right)^9$

11. 0.35 is _____ $\left(\frac{1}{2}\right)^2$

   **A** less than

   **B** greater than

   **C** equal to

   **D** not enough information

12. $5^2$ is _____ $\sqrt[3]{125}$

   **A** less than

   **B** greater than

   **C** equal to

   **D** not enough information

# Chapter 3
# Solving Multi-Step Linear Equations and Inequalities

This chapter covers the following IN Algebra I standards:

| Standard 2: | Linear Equations and Inequalities | A1.2.1 |
| | | A1.2.2 |
| | | A1.2.3 |
| | | A1.2.4 |
| | | A1.2.5 |
| Standard 9: | Mathematical Reasoning and Problem Solving | A1.9.3 |
| | | A1.9.4 |

### 3.1 Two-Step Linear Equations

In the following two-step algebra problems, **additions** and **subtractions** are performed first and then **multiplication** and **division**.

**Example 1:** $\qquad -4x + 7 = 31$

**Step 1:** Subtract 7 from both sides.

$$
\begin{array}{rl}
-4x + 7 & = 31 \\
-7 & \ -7 \\
\hline
-4x & = 24
\end{array}
$$

**Step 2:** Divide both sides by $-4$.

$$\frac{-4x}{-4} = \frac{24}{-4} \qquad \text{so } x = -6$$

**Example 2:** $\qquad -8 - y = 12$

**Step 1:** Add 8 to both sides.

$$
\begin{array}{rl}
-8 - y & = 12 \\
+8 & \ +8 \\
\hline
-y & = 20
\end{array}
$$

**Step 2:** To finish solving a problem with a negative sign in front of the variable, multiply both sides by $-1$. The variable needs to be positive in the answer.

$$(-1)(-y) = (-1)(20) \text{ so } y = -20$$

**Solve the two-step algebra problems below.**

1. $6x - 4 = -34$

2. $5y - 3 = 32$

3. $8 - t = 1$

4. $10p - 6 = -36$

5. $11 - 9m = -70$

6. $4x - 12 = 24$

7. $3x - 17 = -41$

8. $9d - 5 = 49$

9. $10h + 8 = 78$

10. $-6b - 8 = 10$

11. $-g - 24 = -17$

12. $-7k - 12 = 30$

13. $9 - 5r = 64$

14. $6y - 14 = 34$

15. $12f + 15 = 51$

16. $21t + 17 = 80$

17. $20y + 9 = 149$

18. $15p - 27 = 33$

19. $22h + 9 = 97$

20. $-5 + 36w = 175$

## 3.2   Two-Step Linear Equations with Fractions

An algebra problem may contain a fraction. Study the following example to understand how to solve algebra problems that contain a fraction.

**Example 3:** $\dfrac{x}{2} + 4 = 3$

**Step 1:**

$$\begin{array}{r} \dfrac{x}{2} + 4 \quad = 3 \\ \underline{-4 \qquad -4} \\ \dfrac{x}{2} \qquad = -1 \end{array}$$

Subtract 4 from both sides.

**Step 2:** $\dfrac{x}{2} = -1$   Multiply both sides by 2 to eliminate the fraction.

$$\dfrac{x}{2} \times 2 = -1 \times 2, \, x = -2$$

**Simplify the following algebra problems.**

1. $4 + \dfrac{y}{3} = 7$

2. $\dfrac{a}{2} + 5 = 12$

3. $\dfrac{w}{5} - 3 = 6$

4. $\dfrac{x}{9} - 9 = -5$

5. $\dfrac{b}{6} + 2 = -4$

6. $7 + \dfrac{z}{2} = -13$

7. $\dfrac{x}{2} - 7 = 3$

8. $\dfrac{c}{5} + 6 = -2$

9. $3 + \dfrac{x}{11} = 7$

10. $16 + \dfrac{m}{6} = 14$

11. $\dfrac{p}{3} + 5 = -2$

12. $\dfrac{t}{8} + 9 = 3$

13. $\dfrac{v}{7} - 8 = -1$

14. $5 + \dfrac{h}{10} = 8$

15. $\dfrac{k}{7} - 9 = 1$

16. $\dfrac{y}{4} + 13 = 8$

17. $15 + \dfrac{z}{14} = 13$

18. $\dfrac{b}{6} - 9 = -14$

19. $\dfrac{d}{3} + 7 = 12$

20. $10 + \dfrac{b}{6} = 4$

21. $2 + \dfrac{p}{4} = -6$

22. $\dfrac{t}{7} - 9 = -5$

23. $\dfrac{a}{10} - 1 = 3$

24. $\dfrac{a}{8} + 16 = 9$

## 3.3    More Two-Step Linear Equations with Fractions

Study the following example to understand how to solve algebra problems that contain a different type of fraction.

**Example 4:**    $\dfrac{x+2}{4} = 3$    In this example, "$x+2$" is divided by 4, and not just the $x$ or the 2.

**Step 1:**    $\dfrac{x+2}{4} \times 4 = 3 \times 4$    First multiply both sides by 4 to eliminate the fraction.

**Step 2:**    
$$\begin{array}{rl} x+2 & = 12 \\ -2 & \phantom{=} -2 \\ \hline x & = 10 \end{array}$$
    Next, subtract 2 from both sides.

**Solve the following problems.**

1. $\dfrac{x+1}{5} = 4$

2. $\dfrac{z-9}{2} = 7$

3. $\dfrac{b-4}{4} = -5$

4. $\dfrac{y-9}{3} = 7$

5. $\dfrac{d-10}{-2} = 12$

6. $\dfrac{w-10}{-8} = -4$

7. $\dfrac{x-1}{-2} = -5$

8. $\dfrac{c+40}{-5} = -7$

9. $\dfrac{13+h}{2} = 12$

10. $\dfrac{k-10}{3} = 9$

11. $\dfrac{a+11}{-4} = 4$

12. $\dfrac{x-20}{7} = 6$

13. $\dfrac{t+2}{6} = -5$

14. $\dfrac{b+1}{-7} = 2$

15. $\dfrac{f-9}{3} = 8$

16. $\dfrac{4+w}{6} = -6$

17. $\dfrac{3+t}{3} = 10$

18. $\dfrac{x+5}{5} = -3$

19. $\dfrac{g+3}{2} = 11$

20. $\dfrac{k+1}{-6} = 5$

21. $\dfrac{y-14}{2} = -8$

22. $\dfrac{z-4}{-2} = 13$

23. $\dfrac{w+2}{15} = -1$

24. $\dfrac{3+h}{3} = 6$

## 3.4   Combining Like Terms

In algebra problems, separate **terms** by $+$ and $-$ signs. The expression $5x - 4 - 3x + 7$ has 4 terms: $5x$, $4$, $3x$, and $7$. Terms having the same variable can be combined (added or subtracted) to simplify the expression. $5x - 4 - 3x + 7$ simplifies to $2x + 3$.

$$5x - 3x \quad - 4 + 7 = 2x + 3$$

**Simplify the following expressions.**

1. $7x + 12x$

2. $8y - 5y + 8$

3. $4 - 2x + 9$

4. $11a - 16 - a$

5. $9w + 3w + 3$

6. $-5x + x + 2x$

7. $w - 15 + 9w$

8. $21 - 10t + 9 - 2t$

9. $-3 + x - 4x + 9$

10. $7b + 12 + 4b$

11. $4h - h + 2 - 5$

12. $-6k + 10 - 4k$

13. $2a + 12a - 5 + a$

14. $5 + 9c - 10$

15. $-d + 1 + 2d - 4$

16. $-8 + 4h + 1 - h$

17. $12x - 4x + 7$

18. $10 + 3z + z - 5$

19. $14 + 3y - y - 2$

20. $11p - 4 + p$

21. $11m + 2 - m + 1$

## 3.5   Solving Linear Equations with Like Terms

When an equation has two or more like terms on the same side of the equation, combine like terms as the **first** step in solving the equation.

**Example 5:** $\quad 7x + 2x - 7 = 21 + 8$

**Step 1:** Combine like terms on both sides of the equation.

$$\begin{aligned} 7x + 2x - 7 &= 21 + 8 \\ 9x - 7 &= 29 \end{aligned}$$

**Step 2:** Solve the two-step algebra problem as explained previously.

$$\begin{aligned} +7 \qquad &+7 \\ 9x \div 9 &= 36 \div 9 \\ x &= 4 \end{aligned}$$

**Solve the equations below combining like terms first.**

1. $3w - 2w + 4 = 6$

2. $7x + 3 + x = 16 + 3$

3. $5 - 6y + 9y = -15 + 5$

4. $-14 + 7a + 2a = -5$

5. $-2t + 4t - 7 = 9$

6. $9d + d - 3d = 14$

7. $-6c - 4 - 5c = 10 + 8$

8. $15m - 9 - 6m = 9$

9. $-4 - 3x - x = -16$

10. $9 - 12p + 5p = 14 + 2$

11. $10y + 4 - 7y = -17$

12. $-8a - 15 - 4a = 9$

If the equation has like terms on both sides of the equation, you must get all of the terms with a **variable** on one side of the equation and all of the **integers** on the other side of the equation.

**Example 6:**      $3x + 2 = 6x - 1$

| | | |
|---|---|---|
| **Step 1:** | Subtract $6x$ from both sides to move all the **variables** to the left side. | $\begin{aligned} 3x + 2 &= 6x - 1 \\ -6x &\quad -6x \\ \hline -3x + 2 &= -1 \end{aligned}$ |
| **Step 2:** | Subtract 2 from both sides to move all the **integers** to the right side. | $\begin{aligned} -2 &\quad -2 \\ \hline \end{aligned}$ |
| **Step 3:** | Divide by $-3$ to solve for $x$. | $\begin{aligned} \frac{-3x}{-3} &= \frac{-3}{-3} \\ x &= 1 \end{aligned}$ |

**Solve the following problems.**

1. $3a + 1 = a + 9$

2. $2d - 12 = d + 3$

3. $5x + 6 = 14 - 3x$

4. $15 - 4y = 2y - 3$

5. $9w - 7 = 12w - 13$

6. $10b + 19 = 4b - 5$

7. $-7m + 9 = 29 - 2m$

8. $5x - 26 = 13x - 2$

9. $19 - p = 3p - 9$

10. $-7p - 14 = -2p + 11$

11. $16y + 12 = 9y + 33$

12. $13 - 11w = 3 - w$

13. $-17b + 23 = -4 - 8b$

14. $k + 5 = 20 - 2k$

15. $12 + m = 4m + 21$

16. $7p - 30 = p + 6$

17. $19 - 13z = 9 - 12z$

18. $8y - 2 = 4y + 22$

19. $5 + 16w = 6w - 45$

20. $-27 - 7x = 2x + 18$

21. $-12x + 14 = 8x - 46$

22. $27 - 11h = 5 - 9h$

23. $5t + 36 = -6 - 2t$

24. $17y + 42 = 10y + 7$

25. $22x - 24 = 14x - 8$

26. $p - 1 = 4p + 17$

27. $4d + 14 = 3d - 1$

28. $7w - 5 = 8w + 12$

29. $-3y - 2 = 9y + 22$

30. $17 - 9m = m - 23$

## 3.6   Solving for a Variable

Sometimes an equation has two variables and you may be asked to solve for one of the variables.

**Example 7:**      If $5x + y = 19$, then $y =$

**Solution:**      The goal is to have only $y$ on one side of the equation and the rest of the terms on the other side of the equation. Follow order of operations to solve.

$5x + y - 5x = 19 - 5x$      Subtract $5x$ from both sides of the equation.
$y = 19 - 5x$

**Example 8:**      If $7m + n = 30$, then $m =$

**Solution:**      The goal is to have only $m$ on one side of the equation and the rest of the terms on the other side of the equation. Follow order of operations to solve.

$7m + n = 30$      Subtract $n$ from both sides of the equation.

$7m + n - n = 30 - n$

$\dfrac{7m}{7} = \dfrac{30 - n}{7}$      Divide both sides of the equation by 7.

$m = \dfrac{30 - n}{7}$

**Solve each of the equations below for the variable indicated. Be sure to follow order of operations.**

1. If $4a + b = 12$, then $a =$

2. If $6c - d = 17$, then $d =$

3. If $3m - n = 11$, then $m =$

4. If $7r + 5s = 35$, then $r =$

5. If $8m - 9n - 2m = 6n$, then $m =$

6. If $-10y - 3x - x = -12y$, then $x =$

7. If $-4t + 4t - 2s = 8$, then $s =$

8. If $5x - 7y + 9y = -9x + 3$, then $y =$

9. If $-10b + 7a + 3a = -4b$, then $a =$

10. If $7x - 8y + x = 5y + 3$, then $x =$

# 3.7   Removing Parentheses

The distributive principle is used to remove parentheses.

**Example 9:**     $2(a+6)$

You multiply 2 by each term inside the parentheses. $2 \times a = 2a$ and $2 \times 6 = 12$. The 12 is a positive number so use a plus sign between the terms in the answer.
$2(a+6) = 2a+12$

**Example 10:**     $4(-5c+2)$

The first term inside the parentheses could be negative. Multiply in exactly the same way as the examples above. $4 \times (-5c) = -20c$ and $4 \times 2 = 8$
$4(-5c+2) = -20c+8$

**Remove the parentheses in the problems below.**

1. $7(n+6)$
2. $8(2g-5)$
3. $11(5z-2)$
4. $6(-y-4)$
5. $3(-3k+5)$

6. $4(d-8)$
7. $2(-4x+6)$
8. $7(4+6p)$
9. $5(-4w-8)$
10. $6(11x+2)$

11. $10(9-y)$
12. $9(c-9)$
13. $12(-3t+1)$
14. $3(4y+9)$
15. $8(b+3)$

The number in front of the parentheses can also be negative. Remove these parentheses the same way.

**Example 11:**     $-2(b-4)$

First, multiply $-2 \times b = -2b$
Second, multiply $-2 \times -4 = 8$
Copy the two products. The second product is a positive number so put a plus sign between the terms in the answer.
$-2(b-4) = -2b+8$

**Remove the parentheses in the following problems.**

16. $-7(x+2)$
17. $-5(4-y)$
18. $-4(2b-2)$
19. $-2(8c+6)$
20. $-5(-w-8)$

21. $-3(4x-2)$
22. $-2(-z+2)$
23. $-4(7p+7)$
24. $-9(t-6)$
25. $-10(2w+4)$

26. $-3(9-7p)$
27. $-9(-k-3)$
28. $-1(7b-9)$
29. $-6(-5t-2)$
30. $-7(-v+4)$

## 3.8 Multi-Step Linear Equations

You can now use what you know about removing parentheses, combining like terms, and solving simple algebra problems to solve problems that involve three or more steps. Study the examples below to see how easy it is to solve multi-step problems.

**Example 12:** $3(x+6) = 5x - 2$

| | | |
|---|---|---|
| **Step 1:** | Use the distributive property to remove parentheses. | $3x + 18 = 5x - 2$ |
| **Step 2:** | Subtract $5x$ from each side to move the terms with variables to the left side of the equation. | $\dfrac{-5x \qquad -5x}{-2x + 18 = -2}$ |
| **Step 3:** | Subtract 18 from each side to move the integers to the right side of the equation. | $\dfrac{-18 \qquad -18}{\dfrac{-2x}{-2} = \dfrac{-20}{-2}}$ |
| **Step 4:** | Divide both sides by $-2$ to solve for $x$. | $x = 10$ |

**Example 13:** $\dfrac{3(x-3)}{2} = 9$

| | | |
|---|---|---|
| **Step 1:** | Use the distributive property to remove parentheses. | $\dfrac{3x - 9}{2} = 9$ |
| **Step 2:** | Multiply both sides by 2 to eliminate the fraction. | $\dfrac{2(3x-9)}{2} = 2(9)$ |
| **Step 3:** | Add 9 to both sides, and combine like terms. | $\dfrac{3x - 9 = 18}{+9 \qquad +9}$ |
| **Step 4:** | Divide both sides by 3 to solve for $x$. | $\dfrac{3x}{3} = \dfrac{27}{3}$ |
| | | $x = 9$ |

**Solve the following multi-step algebra problems.**

1. $2(y-3) = 4y + 6$

2. $\dfrac{2(a+4)}{2} = 12$

3. $\dfrac{10(x-2)}{5} = 14$

4. $\dfrac{12y - 18}{6} = 4y + 3$

5. $2x + 3x = 30 - x$

6. $\dfrac{2a+1}{3} = a + 5$

7. $5(b-4) = 10b + 5$

8. $-8(y+4) = 10y + 4$

9. $\dfrac{x+4}{-3} = 6 - x$

10. $\dfrac{4(n+3)}{5} = n - 3$

11. $3(2x - 5) = 8x - 9$

12. $7 - 10a = 9 - 9a$

13. $7 - 5x = 10 - (6x + 7)$

14. $4(x - 3) - x = x - 6$

15. $4a + 4 = 3a - 4$

16. $-3(x - 4) + 5 = -2x - 2$

17. $5b - 11 = 13 - b$

18. $\dfrac{-4x + 3}{2x} = \dfrac{7}{2x}$

19. $-(x + 1) = -2(5 - x)$

20. $4(2c + 3) - 7 = 13$

21. $6 - 3a = 9 - 2(2a + 5)$

22. $-5x + 9 = -3x + 11$

23. $3y + 2 - 2y - 5 = 4y + 3$

24. $3y - 10 = 4 - 4y$

25. $-(a + 3) = -2(2a + 1) - 7$

26. $5m - 2(m + 1) = m - 10$

27. $\dfrac{1}{2}(b - 2) = 5$

28. $-3(b - 4) = -2b$

29. $4x + 12 = -2(x + 3)$

30. $\dfrac{7x + 4}{3} = 2x - 1$

31. $9x - 5 = 8x - 7$

32. $7x - 5 = 4x + 10$

33. $\dfrac{4x + 8}{2} = 6$

34. $2(c + 4) + 8 = 10$

35. $y - (y + 3) = y + 6$

36. $4 + x - 2(x - 6) = 8$

## 3.9   Solving Linear Inequalities Using Solution Sets

An **inequality** is a sentence that contains a $\neq$, $<$, $>$, $\leq$, or $\geq$ sign. When these objects are numbers like 9 and 5, we can truthfully say that $9 > 5$, "9 is greater than 5", and $5 < 9$, "5 is less than 9". Variables and coefficients might also come into play in statements of inequality such as

$$5x < 25.$$

Here, the statement will be true or false based on the values that $x$ takes. For the set, $\{3, 4, 5, 6\}$, the statement is true for $x = 3$ or 4 but false for $x = 5$ or 6. 15 and 20 are less than 25 but 25 and 30 are not less than 25. Solving inequalities is much like solving equations. In the above inequality we can divide both sides by 5 and get $x < 5$. This statement is true for all values of $x$ that are less than 5. Observe a crucial difference between equations and inequalities:

$$-x < 5 \iff x > -5$$

When dividing or multiplying by a negative number, switch the direction of the inequality sign. So, if it is true that $-x < 5$, it is also true that $x > -5$. Try plugging integers greater than $-5$ such as $\{-4, -3, -2, -1, 0, \ldots\}$ into the original inequality.

**Example 14:**   From the set $\{0, 1, 2, 3\}$, find the solution set to the inequality $5x + 2 > 10$.

**Step 1:**   In the inequality, solve for $x$.
Subtracting 2 from each side leaves $5x > 8$.

**Step 2:**   Dividing each side by 5 gives $x > \frac{8}{5}$.

**Step 3:**   Find the numbers in the given set that are greater than $\frac{8}{5}$. Since $\frac{8}{5}$ is greater than 1 but less than 2, we can tell that $\{2, 3\}$ make the original inequality statement true whereas 0 and 1 do not.

**Step 4:**   Check your work. It's easy to see that $5(2) = 10 > 8$ and $5(3) = 15 > 8$ but $5(1) = 5 \not> 8$ and $5(0) = 0 \not> 8$. Thus, from the given set, the solution set is $\{2, 3\}$.

**Example 15:**   Solve the inequality $-x + 1 > 0$ for $x$ in the set $\{-53, -7, 11, 22, 51\}$.

**Step 1:**   In the inequality, solve for $x$.
Subtract 1 from both sides. $-x > -1$

**Step 2:**   Multiply both sides by $-1$. Don't forget to change the direction the inequality symbol. Now, $x < 1$.

**Step 3:**   Find the values from the given set that make the inequality true. In this case the values from our set $\{-53, -7, 11, 22, 51\}$ that are less than 1 happen to be $\{-53, -7\}$.
$\{-53, -7\}$ is the solution set.

**From the given set and inequality, find the solution set.**

1. $5x > 100$ in $\{0, 10, 20, 30, 40\}$

2. $7x - 10 < 30$ in $\{4, 5, 6, 7\}$

3. $-10x + 5 < 0$ in $\{-1, 0, 1, 2\}$

4. $3x + 7 > 28$ in $\{5, 6, 7, 8, 9, 10\}$

5. $-x - 10 > 0$ in $\{-30, -20, -10, 0, 10, 20\}$

6. $\frac{1}{5}x > 5$ in $\{-5, 0, 10, 25, 45\}$

7. $1 - x < 5$ in $\{-97, -60, 97, -1, -70, -47\}$

8. $-7x - 7 < -10$ in $\{-45, 21, 49, -12, 8, 45\}$

9. $5x - 8 < -4$ in $\{-63, -47, -48, -99, 11, 76\}$

10. $-x - 2 < 3$ in $\{-1, -2, -2, 1, -5, 0\}$

11. $3x + 4 < -4$ in $\{-5, -4, -3, -2, -1, 0\}$

12. $-x - 4 < 2$ in $\{26, -29, 1, -19, -26, -5\}$

## 3.10   Multi-Step Linear Inequalities

Remember that adding and subtracting with inequalities follow the same rules as equations. When you multiply or divide both sides of an inequality by the same positive number, the rules are the same as for equations. However, when you multiply or divide both sides of an inequality by a **negative** number, you must **reverse** the inequality symbol.

**Example 16:**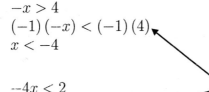
$$-x > 4$$
$$(-1)(-x) < (-1)(4)$$
$$x < -4$$

**Example 17:**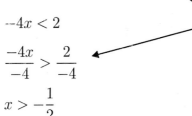
$$-4x < 2$$
$$\frac{-4x}{-4} > \frac{2}{-4}$$
$$x > -\frac{1}{2}$$

Reverse the symbol when you multiply or divide by a negative number.

When solving multi-step inequalities, first add and subtract to isolate the term with the variable. Then multiply and divide.

**Example 18:**   $2x - 8 > 4x + 1$

**Step 1:**   Add 8 to both sides.
$$2x - 8 + 8 > 4x + 1 + 8$$
$$2x > 4x + 9$$

**Step 2:**   Subtract $4x$ from both sides.
$$2x - 4x > 4x + 9 - 4x$$
$$-2x > 9$$

**Step 3:**   Divide by $-2$. Remember to change the direction of the inequality sign.
$$\frac{-2x}{-2} < \frac{9}{-2}$$
$$x < -\frac{9}{2}$$

## Solve each of the following inequalities.

1. $8 - 3x \leq 7x - 2$

2. $3(2x - 5) \geq 8x - 5$

3. $\frac{1}{3}b - 2 > 5$

4. $7 + 3y > 2y - 5$

5. $3a + 5 < 2a - 6$

6. $3(a - 2) > -5a - 2(3 - a)$

7. $2x - 7 \geq 4(x - 3) + 3x$

8. $6x - 2 \leq 5x + 5$

9. $-\frac{x}{4} > 12$

10. $-\frac{2x}{3} \leq 6$

11. $3b + 5 < 2b - 8$

12. $4x - 5 \leq 7x + 13$

13. $4x + 5 \leq -2$

14. $2y - 5 > 7$

15. $4 + 2(3 - 2y) \leq 6y - 20$

16. $-4c + 6 \leq 8$

17. $-\frac{1}{2}x + 2 > 9$

18. $\frac{1}{4}y - 3 \leq 1$

19. $-3x + 4 > 5$

20. $\frac{y}{2} - 2 \geq 10$

21. $7 + 4c < -2$

22. $2 - \frac{a}{2} > 1$

23. $10 + 4b \leq -2$

24. $-\frac{1}{2}x + 3 > 4$

## 3.11    Combined Linear Inequalities

Inequalities can have more than one inequality sign. These are called **combined inequalities**. For example:

$-2 \leq x < 4$ is read "$-2$ is less than or equal to $x$ and $x$ is less than 4."

You learned earlier in this chapter that you can perform any operation on an equation, as long as you perform the same operation to the other side. The same is true of inequalities. It's also true of combined inequalities.

**Example 19:**    Solve the combined inequality $-20 < 5x - 10 < 20$ for $x$.

**Step 1:**    We must isolate $x$. The first step is to add 10 to each part of the combined inequality.

$$
\begin{array}{ccccc}
-20 & < & 5x - 10 & < & 20 \\
+10 & & +10 & & +10 \\
\hline
-10 & < & 5x & < & 30
\end{array}
$$

**Step 2:**    Now divide each part by 5.

$$\frac{-10}{5} < \frac{5x}{5} < \frac{30}{5}$$

$$-2 < x < 6$$

**Solve the inequalities.**

1. $-14 < 2x + 4 \leq 14$

2. $6 > 4b - 8 > 0$

3. $-36 < 12e + 3 < 36$

4. $49 > 8f - 15 > 1$

5. $-98 < 7g - 91 < 0$

6. $14 > 4a + 10 \geq -14$

7. $13 \leq 9k - 32 < 121$

8. $35 > -5j - 50 > -10$

9. $43 > 6f - 83 > 19$

10. $43 > 31 + 6x \geq 4$

11. $7 > 9d - 8 > -10$

12. $48 > 10c - 12 > 23$

13. $-6 < 29 + 7v < 15$

14. $-2 < 9e + 7 \leq 11$

15. $77 \geq 8a + 21 > 29$

16. $-4 \leq 3g - 40 \leq 38$

17. $2 > 10 - 8r > 34$

18. $37 \geq 11u - 29 \geq 4$

19. $35 > 9q - 10 > -28$

20. $36 \geq -6v - 18 > 18$

# Chapter 3 Review

**Solve each of the following equations.**

1. $4a - 8 = 28$

2. $5 + \dfrac{x}{8} = -4$

3. $-7 + 23w = 108$

4. $\dfrac{y - 8}{6} = 7$

5. $c - 13 = 5$

6. $\dfrac{b + 9}{12} = -3$

**Simplify the following expressions by combining like terms.**

7. $-4a + 8 + 3a - 9$

8. $14 + 2z - 8 - 5z$

9. $-7 - 7x - 2 - 9x$

**Solve.**

10. $19 - 8d = d - 17$

11. $7w - 8w = -4w - 30$

12. $6 + 16x = -2x - 12$

13. $6(b - 4) = 8b - 18$

14. $4x - 16 = 7x + 2$

15. $9w - 2 = -w - 22$

**Remove parentheses.**

16. $3(-4x + 7)$

17. $11(2y + 5)$

18. $6(8 - 9b)$

19. $-8(-2 + 3a)$

20. $-2(5c - 3)$

21. $-5(7y - 1)$

**Solve for the variable.**

22. If $3x - y = 15$, then $y =$

23. If $7a + 2b = 1$, then $b =$

**Solve each of the following equations and inequalities.**

24. $\dfrac{-11c - 35}{4} = 4c - 2$

25. $5 + x - 3(x + 4) = -17$

26. $4(2x + 3) \geq 2x$

27. $7 - 3x \leq 6x - 2$

28. $\dfrac{5(n + 4)}{3} = n - 8$

29. $-y > 14$

30. $2(3x - 1) \geq 3x - 7$

31. $3(x + 2) < 7x - 10$

32. $-22 < -11k < 22$

33. $18 > -3w > -18$

34. $21 \leq 5x + 6 \leq 46$

35. $20 > 3x + 2 \geq 1$

**From the given set and inequality, find the solution set.**

36. $10x < 3$ in $\{-2, -1, 0, 1, 2\}$

37. $-2x + 9 > 5$ in $\{0, 1, 2, 3, 4\}$

# Chapter 3 Test

1. Find the value of $n$. $19n - 57 = 76$

   **A** 1
   **B** 3
   **C** 5
   **D** 7

2. Solve for $x$. $14x + 84 = 154$

   **A** 4
   **B** 5
   **C** 11
   **D** 17

3. Which of the following is equivalent to $4 - 5x > 3(x - 4)$?

   **A** $4 - 5x > 3x - 4$
   **B** $4 - 5x > 3x - 12$
   **C** $4 - 5x > 3x - 1$
   **D** $4 - 5x > 3x - 7$

4. Which of the following is equivalent to $3(x - 2) + 1 - 2x = -4$?

   **A** $x - 6 = -4$
   **B** $-6x + 1 = -4$
   **C** $5x - 7 = -4$
   **D** $x - 5 = -4$

5. Solve: $3(x - 2) - 1 = 6(x + 5)$

   **A** $-4$
   **B** $-\dfrac{37}{3}$
   **C** $4$
   **D** $\dfrac{23}{3}$

6. $5(2x + 11) - 3 \times 5 = ?$

   **A** $7x + 40$
   **B** $7x + 20$
   **C** $10x + 40$
   **D** $10x + 260$

7. Solve: $4b - 8 < 56$

   **A** $b < 12$
   **B** $b < 16$
   **C** $b < -12$
   **D** $b < -16$

8. Which of the following is equivalent to $3(2x - 5) - 4(x - 3) = 7$?

   **A** $x + 27 = 7$
   **B** $2x - 3 = 7$
   **C** $10x - 27 = 7$
   **D** $x - 27 = 7$

9. Which of the following is true of the inequality $-\dfrac{x}{3} + 4 < 4$?

   **A** The inequality is true for all positive values of $x$.
   **B** The inequality is true for all negative values of $x$.
   **C** The inequality is true for all non-zero values of $x$.
   **D** The inequality is false for all positive values of $x$.

10. $-x + 2 < 10$ is true for

   **A** All values of $x$.
   **B** No values of $x$.
   **C** $x > -8$.
   **D** $x < 8$.

# Chapter 4
# Algebra Word Problems

This chapter covers the following IN Algebra I standards:

| Standard 1: | Operations with Real Numbers | A1.1.5 |
| Standard 2: | Linear Equations and Inequalities | A1.2.6 |

## 4.1   Rate

**Example 1:**   Laurie traveled 312 miles in 6 hours. What was her average rate of speed?

Divide the number of miles by the number of hours.   $\dfrac{312 \text{ miles}}{6 \text{ hours}} = 52 \text{ miles/hour}$

**Laurie's average rate of speed was 52 miles per hour (or 52 mph).**

**Find the average rate of speed in each problem below.**

1. A race car went 500 miles in 4 hours. What was its average rate of speed?

2. Carrie drove 124 miles in 2 hours. What was her average speed?

3. After 7 hours of driving, Chad had gone 364 miles. What was his average speed?

4. Anna drove 360 miles in 8 hours. What was her average speed?

5. After 3 hours of driving, Paul had gone 183 miles. What was his average speed?

6. Nicole ran 25 miles in 5 hours. What was her average speed?

7. A train traveled 492 miles in 6 hours. What was its average rate of speed?

8. A commercial jet traveled 1,572 miles in 3 hours. What was its average speed?

9. Jillian drove 195 miles in 3 hours. What was her average speed?

10. Greg drove 8 hours from his home to a city 336 miles away. At what average speed did he travel?

11. Caleb drove 128 miles in two hours. What was his average speed in miles per hour?

12. After 9 hours of driving, Kate had traveled 405 miles. What speed did she average?

## 4.2   More Rates

Rates are often discussed in terms of miles per hour, but a rate can be any measured quantity divided by another measurement such as feet per second, kilometers per minute, mass per unit volume, etc. A rate can be how fast something is done. For example, a bricklayer may lay 80 bricks per hour. Rates can also be used to find measurements such as density. For example, 35 grams of salt in 1 liter of water gives the mixture a density of 35 grams/liter.

**Example 2:**   Nathan entered his snail in a race. His snail went 18 feet in 6 minutes. How fast did his snail move?

In this problem, the units given are feet and minutes, so the rate will be feet per minute (or feet/minute).

You need to find out how far the snail went in one minute.

$$\text{Rate equals } \frac{\text{distance}}{\text{time}} \text{ so } \frac{18 \text{ feet}}{6 \text{ minutes}} = \frac{3 \text{ feet}}{1 \text{ minute}}$$

**Nathan's snail went an average of 3 feet per minute or $3\dfrac{\text{ft}}{\text{min}}$.**

**Find the average rate for each of the following problems.**

1. Tewanda read a 2,000-word news article in 8 minutes. How fast did she read the news article?

2. Chandler rides his bike to school every day. He travels 2,560 feet in 640 seconds. How many feet did he travel per second?

3. Mr. Molier is figuring out the semester averages for his history students. He can calculate the average for 20 students in an hour. How long does it take him to figure the average for each student?

4. In 1908, John Hurlinger of Austria walked 1,400 kilometers from Vienna to Paris on his hands. The journey took 55 days. What was his average speed per day?

5. Spectators at the Super Circus were amazed to watch a cannon shoot a clown 212 feet into a net in 4 seconds. How many feet per second did the clown travel?

6. Marcus Page, star receiver for the Big Bulls, was awarded a 5-year contract for 105 million dollars. How much will his annual rate of pay be if he is paid the same amount each year?

7. Duke Delaney scored 28 points during the 4 quarters of the basketball playoffs. What was his average score per quarter?

8. The new McDonald's in Moscow serves 11,208 customers during a 24-hour period. What is the average number of customers served per hour?

## 4.3   Dimensional Analysis

A **dimension** is a property that can be measured such as length, time, mass, or temperature; or it is calculated by multiplying or dividing other dimensions. Some examples include length/time (velocity), length$^3$ (volume), or mass/length$^3$ (density).

A measured quantity can be expressed in any appropriate dimension. The equivalence between two expressions of a given quantity may be written as a ratio:

$$\frac{16 \text{ ounces}}{1 \text{ pound}} \quad \text{or} \quad \frac{2000 \text{ pounds}}{1 \text{ ton}}$$

Ratios of equivalent values expressed in different units like these are known as **conversion factors**. To convert given quantities in one set of units to their equivalent values in another set of units, we set up **dimensional equations**. Writing these with vertical and horizontal bars and carrying along units often helps avoid mistakes in these types of equations.

**Example 3:**   How many inches are in four yards?

**Step 1:**   First we need to develop our conversion factors. We are given a value in yards and we want to find the value in inches. We know that there are 3 feet in 1 yard, and that there are 12 inches in 1 foot, so we will set up a dimensional equation using these conversion factors. We will then be able to cross out the units top and bottom to make sure that we are left with the units that we want. Then we will multiply the numbers across the top and divide by the numbers on the bottom to get the answer.

$$\frac{4 \text{ yards}}{} \left| \frac{3 \text{ feet}}{1 \text{ yard}} \right| \frac{12 \text{ inches}}{1 \text{ foot}} = \frac{(4) \times (3) \times (12)}{(1) \times (1)} = 144 \text{ inches}$$

**Example 4:**   How many second are there in March?

$$\frac{60 \text{ seconds}}{1 \text{ minute}} \left| \frac{60 \text{ minutes}}{1 \text{ hour}} \right| \frac{24 \text{ hours}}{1 \text{ day}} \left| \frac{31 \text{ days}}{\text{March}} \right. = 2{,}678{,}400 \text{ seconds/March}$$

We are often asked to convert the units of area or volume, which are square or cubic terms. It is important to note that the numerical coefficient must also be squared or cubed when converting units that are squared or cubed.

**Example 5:**   The volume of a barrel is 10 ft$^3$. How many in$^3$ of water will it hold?

$$\frac{10 \text{ ft}^3}{} \left| \frac{(12)^3 \text{ in}^3}{1 \text{ ft}^3} \right. = \frac{(10) \times (12) \times (12) \times (12)}{(1)} = 17{,}280 \text{ in}^3$$

**Example 6:**   How many kilograms per cubic meter ($kg/m^3$) are there in 3 grams per cubic centimeter ($g/cm^3$)?

$$\frac{3 \text{ grams}}{cm^3} \left| \frac{1 \text{ kg}}{1000 \text{ grams}} \right| \frac{(100)^3 cm^3}{1 m^3} = 3000 \frac{kg}{m^3}$$

This is an example of converting units of **density**, which is a measure in mass per unit volume. Another useful scientific term which often needs converting is **velocity**, measured in length per unit time.

**Example 7:**   What is the highway speed limit of 65 miles per hour in feet per second?

$$\frac{65 \text{ miles}}{1 \text{ hour}} \left| \frac{5,280 \text{ feet}}{1 \text{ mile}} \right| \frac{1 \text{ hour}}{60 \text{ minutes}} \left| \frac{1 \text{ minute}}{60 \text{ seconds}} \right. = 95.3 \frac{\text{feet}}{\text{second}}$$

**Problems.**

1. Jared can work 54 math problems in one hour. How many problems can he work in 10 minutes?

2. Leah rides 22 feet per second on her bicycle. How many miles per hour does she ride?

3. A jar of honey has a density of 14 kg per $m^3$. What is its density in $g/mm^3$?

4. If a pitcher will hold 2 $ft^3$ of lemonade, how many $in^3$ of lemonade will it hold?

5. Juan's car gets an average of 24 miles per gallon of gas. How far can Juan go on 1 quart of gas?

6. From question 5, how many gallons of gas will it take Juan to travel 528 miles?

7. How many cubic centimeters ($cm^3$) of water are in 1 $m^3$?

8. How many cubic feet are in a hole that is 3 feet deep, 4 feet wide, and 6 feet long?

9. How many yards are in 3 miles?

10. John Smoltz throws his fastball 99 miles per hour. If he starts 60 feet, 6 inches from home plate, how many seconds does it take the ball to get to the plate?

## 4.4   Algebra Word Problems

An equation states that two mathematical expressions are equal. In working with word problems, the words that mean equal are **equals, is, was, is equal to, amounts to,** and other expressions with the same meaning. To translate a word problem into an algebraic equation, use a variable to represent the unknown or unknowns you are looking for.

In the following example, let $n$ be the number you are looking for.

**Example 8:**   Four more than twice a number is two less than three times the number.

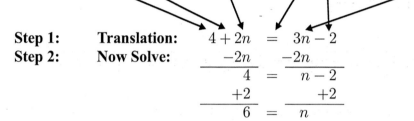

**Step 1:**   **Translation:**   $4 + 2n = 3n - 2$
**Step 2:**   **Now Solve:**

$$\begin{array}{rcl} 4 + 2n & = & 3n - 2 \\ -2n & & -2n \\ \hline 4 & = & n - 2 \\ +2 & & +2 \\ \hline 6 & = & n \end{array}$$

The number is 6.
Substitute the number back into the original equation to check.

**Translate the following word problems into equations and solve.**

1. Four less than twice a number is ten. Find the number.

2. Three more than three times a number is one less than two times the number. What is the number?

3. The sum of seven times a number and the number is 24. What is the number?

4. Negative 18 is the sum of five and a number. Find the number.

5. Negative 14 is equal to ten minus the product of six and a number. What is the number?

6. Two less than twice a number equals the number plus 12. What is the number?

7. The difference between three times a number and 31 is two. What is the number?

8. Sixteen is fourteen less than the product of a number and five. What is the number?

9. Eight more than twice a number is four times the difference between five and the number. What is the number?

10. Three less than twice a number is three times the sum of one and the number. What is the number?

## 4.5   Real-World Linear Equations

Linear equations are very useful mathematical tools. They allow us to show relationships between two variables.

**Example 9:**   A local cell phone company uses the equation $y = \frac{5}{2}x + 10$ to determine the charges for usage where $y = $ the cost and $x = $ the minutes used. How much will Jessica's bill be if she talked for 40 minutes?

**Step 1:**   Substitute the known value in for $x$.
$y = \frac{5}{2}(40) + 10$

**Step 2:**   Simplify.
$y = 100 + 10 = 110$
Jessica's bill will be $110.

**Example 10:**   Vincent bought a luxury car for $165,000$ and its value has depreciated linearly. After 5 years the value was $137,000$. What is the amount of yearly depreciation?

**Step 1:**   First find how much the car's value depreciated in 5 years.
$\$165,000 - \$137,000 = \$28,000$

**Step 2:**   Next, find the yearly depreciation by dividing $\$28,000$ by the amount of years, 5.
$\$28,000 \div 5 = \$5,600$
The value of Vincent's car depreciated $\$5,600$ each year.

**Example 11:**   In 1990, the average cost of a new house was $123,000$. By the year 2000, the average cost of a new house was $134,150$. Based on a linear model, what is the predicted average cost for 2008?

**Step 1:**   First, we need to find the difference between the average cost of a new house in the year 1990 and the average cost of a new house in the year 2000.
$\$134,150 - \$123,000 = \$11,150$

**Step 2:**   Next, we need to find how much the average cost of a new house went up each year. Since it had been 10 years, divide the difference between the value in 2000 and 1990 by 10.
$\$11,150 \div 10 = \$1,115$

**Step 3:**   Multiply the amount the average cost of a new house went up each year by the number of years between 2000 and 2008.
$\$1,115 \times 8 = \$8,920$

**Step 4:**   Lastly, add the average cost of a new house in the year 2000 with the amount found in step 3.
$\$134,150 + \$8,920 = 143,070$
$\$143,070$ is the predicted average cost of a new house for 2008.

**Solve the following problems.**

1. Acacia bought an MP3 player at Everywhere Electronics for $350 and its valued depreciated linearly. Three years later, she saw the same MP3 player at Everywhere Electronics for $125. What is the amount of yearly depreciation of Acacia's MP3 player?

2. Dustin bought a boat 10 years ago for $10,000. Its value depreciated linearly and now it is worth $2,500. What is the amount of yearly depreciation of Dustin's boat?

3. A small plane costs $500,000 new. Twenty years later it is valued at $150,000. Assuming a linear depreciation, what was the value of the plane when it was 14 years old?

4. In 1980, the price of a scientific calculator was $155. In 2005, the price was $15 dollars. Assuming the change in price was linear, what was the price of a scientific calculator in 1997?

5. In 1997, Justin bought a house for $120,000. In 2004, his house was worth $176,000. Based on a linear model, how much was Justin's house worth in 2001?

6. The attendance on the first day of the Sunny Day Festival was 325 people. The attendance on the third day was 382 people. Assuming the attendance will increase linearly each day, how many people will attend the Sunny Day Festival on the seventh day?

7. Two years ago Juanita bought 2 shirts for $15 and last year she bought 4 shirts for $45. Assuming the price will increase linearly, how much will 8 shirts cost Juanita this year?

8. In 1985, the average price of a new car was $9,000. In 2000, the average price was $24,750. Based on a linear model, what is the predicted average price for 2009?

**Use the following information for questions 9–10.**

Abbey is looking for a new cell phone provider. In her search, she has found 3 local companies: Gift of Gab, On the Go, and Connect. To determine their monthly charges, the 3 companies use the following equations.

Gift of Gab: $y = \frac{3}{4}x + 20$

On the Go: $y = \frac{1}{2}x + 60$

Connect: $y = 6x - 100$

9. Which is the cheapest provider if Abbey uses 200 minutes per month?

10. What if she used 100 minutes?

# 4.6    Word Problems with Formulas

The perimeter of a geometric figure is the distance around the outside of the figure.

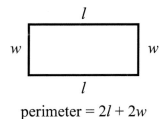

perimeter = $2l + 2w$

perimeter = $a + b + c$

**Example 12:**   The perimeter of a rectangle is 44 feet. The length of the rectangle is 6 feet more than the width. What is the measure of the width?

**Step 1:**   Let the variable be the length of the unknown side.
width = $w$          length = $6 + w$

**Step 2:**   Use the equation for the perimeter of a rectangle as follows:
$2l + 2w$ = perimeter of a rectangle
$2(w + 6) + 2w = 44$

**Step 3:**   Solve for $w$.

**Solution:**   width = 8 feet

**Example 13:**   The perimeter of a triangle is 26 feet. The second side is twice as long as the first. The third side is 1 foot longer than the second side. What is the length of the 3 sides?

**Step 1:**   Let $x$ = first side      $2x$ = second side      $2x + 1$ = third side

**Step 2:**   Use the equation for perimeter of a triangle as follows:
sum of the length of the sides = perimeter of a triangle.
$x + 2x + 2x + 1 = 26$

**Step 3:**   Solve for $x$. $5x + 1 = 26$ so $x = 5$

**Solution:**   first side $x = 5$      second side $2x = 10$      third side $2x + 1 = 11$

**Solve the following word problems.**

1. The length of a rectangle is 4 times longer than the width. The perimeter is 30. What is the width?

2. The length of a rectangle is 3 more than twice the width. The perimeter is 36. What is the length?

3. The perimeter of a triangle is 18 feet. The second side is two feet longer than the first. The third side is two feet longer then the second. What are the lengths of the sides?

4. In an isosceles triangle, two sides are equal. The third side is two less than twice the length of the sum of the two sides. The perimeter is 40. What are the lengths of the three sides?

5. The sum of the measures of the angles of a triangle is 180°. The second angle is three times the measure of the first angle. The third angle is four times the measure of the second angle. Find the measure of each angle.

6. The sum of the measures of the angles of a triangle is 180°. The second angle of a triangle is twice the measure of the first angle. The third angle is 20 more than 5 times the first. What are the measures of the three angles?

## 4.7    Age Problems

**Example 14:**    Tara is twice as old as Gwen. Their sister, Amy, is 5 years older than Gwen. If the sum of their ages is 29 years, find each of their ages.

**Step 1:**    We want to find each of their ages so there are three unknowns. Tara is twice as old as Gwen, and Amy is older than Gwen, so Gwen is the youngest. Let $x$ be Gwen's age. From the problem we can see that:

$$\left. \begin{array}{rcl} \text{Gwen} & = & x \\ \text{Tara} & = & 2x \\ \text{Amy} & = & x + 5 \end{array} \right\} \text{The sum of their ages is 29.}$$

**Step 2:**    Set up the equation, and solve for $x$.

$$\begin{array}{rcl} x + 2x + x + 5 & = & 29 \\ 4x + 5 & = & 29 \\ 4x & = & 29 - 5 \\ x & = & \dfrac{24}{4} \\ x & = & 6 \end{array}$$

**Solution:**    
Gwen's age $(x)$ $=$ 6
Tara's age $(2x)$ $=$ 12
Amy's age $(x + 5) = 11$

**Solve the following age problems.**

1. Carol is 15 years older than her cousin Amanda. Cousin Bill is 4 times as old as Amanda. The sum of their ages is 99. Find each of their ages.

2. Derrick is 5 less than twice as old as Brandon. The sum of their ages is 31. How old are Derrick and Brandon?

3. Beth's mom is 5 times older than Beth. Beth's dad is 8 years older than Beth's mom. The sum of their ages is 74. How old are each of them?

4. Delores is 4 years more than three times as old as her son, Raul. If the difference between their ages is 34, how old are Delores and Raul?

5. Eileen is 9 years older than Karen. John is three times as old as Karen. The sum of their ages is 64. How old are Eileen, Karen, and John?

6. Taylor is 20 years younger than Jim. Andrew is twice as old as Taylor. The sum of their ages is 32. How old are Taylor, Jim, and Andrew?

**The following problems work in the same way as the age problems. There are two or three items of different weight, distance, number, or size. You are given the total and asked to find the amount of each item.**

7. Three boxes have a total height of 720 pounds. Box A weighs twice as much as Box B. Box C weighs 30 pounds more than Box A. How much do each of the boxes weigh?

8. There are 170 students registered for American History classes. There are twice as many students registered in second period as first period. There are 10 less than three times as many students registered in third period as in first period. How many students are in each period?

9. Mei earns $4 less than three times as much as Olivia. Shane earns twice as much as Mei. Together they earn $468 per week. How much does each person earn per week?

10. Ellie, the elephant, eats 4 times as much as Popcorn, the pony. Zac, the zebra, eats twice as much as Popcorn. Altogether, they eat 238 kilograms of feed per week. How much feed does each of them require each week?

11. The school cafeteria served three kinds of lunches today to 117 students. The students chose the cheeseburgers three times more often than the grilled cheese sandwiches. There were twice as many grilled cheese sandwiches sold as fish sandwiches. How many of each lunch were served?

12. Three friends drove southeast to New Mexico. Kyle drove half as far as Jamaal. Conner drove 4 times as far as Kyle. Altogether, they drove 476 miles. How far did each friend drive?

13. Bianca is taking collections for this year's Feed the Hungry Project. She has collected $300 more from Company A than from Company B and $700 more from Company C than from Company A. So far, she has collected $4,300. How much did Company C give?

14. For his birthday, Torin got $50.00 more from his grandmother than from his uncle. His uncle gave him $10.00 less than his cousin. Torin received $135.00 in total. How much did he receive from his cousin?

15. Cassidy loves black and yellow jelly beans. She noticed when she was counting them that she had 7 less than three times as many black jelly beans as she had yellow jelly beans. In total, she counted 225 jelly beans. How many black jelly beans did she have?

16. Mrs. Vargus planted a garden with red and white rose bushes. Because she was studying to be a botanist, she counted the number of blossoms on each bush. She counted 4 times as many red blossoms as white blossoms. In total, she counted 1,420 blossoms. How many red blossoms did she count?

## 4.8   Consecutive Integer Problems

|  | Examples: | Algebraic notation: |
|---|---|---|
| **Consecutive integers** follow each other in order | $1, 2, 3, 4$ <br> $-3, -4, -5, -6$ | $n, n+1, n+2, n+3$ |
| Consecutive **even** integers: | $2, 4, 6, 8, 10$ <br> $-12, -14, -16, -18$ | $n, n+2, n+4, n+6$ |
| Consecutive **odd** integers: | $3, 5, 7, 9$ <br> $-5, -7, -9, -11$ | $n, n+2, n+4, n+6$ |

**Example 15:**   The sum of three consecutive odd integers is 63. Find the integers.

**Step 1:**   Represent the three odd integers:
Let $n$ = the first odd integer
$n + 2$ = the second odd integer
$n + 4$ = the third odd integer

**Step 2:**   The sum of the integers is 63, so the algebraic equation is
$n + n + 2 + n + 4 = 63$.   Solve for $n$.
$n = 19$

**Solution:**   the first odd integer $= 19$
the second odd integer $= 21$
the third odd integer $= 23$

**Check:**   Does $19 + 21 + 23 = 63$? Yes, it does.

**Solve the following problems.**

1. Find three consecutive even integers whose sum is 120.

2. Find three consecutive integers whose sum is $-30$.

3. The sum of three consecutive odd integers is 51. What are the numbers?

4. Find two consecutive odd integers such that five times the first equals three times the second.

5. Find two consecutive even integers such that seven times the first equals six times the second.

6. Find two consecutive odd numbers whose sum is eighty.

## 4.9 Inequality Word Problems

Inequality word problems involve staying under a limit or having a minimum goal one must meet.

**Example 16:** A contestant on a popular game show must earn a minimum of 800 points by answering a series of questions worth 40 points each per category in order to win the game. The contestant will answer questions from each of four categories. Her results for the first three categories are as follows: 160 points, 200 points, and 240 points. Write an inequality which describes how many points, $(p)$, the contestant will need on the last category in order to win.

**Step 1:** Add to find out how many points she already has. $160 + 200 + 240 = 600$

**Step 2:** Subtract the points she already has from the minimum points she needs. $800 - 600 = 200$. She must get at least 200 points in the last category to win. If she gets more than 200 points, that is okay, too. To express the number of points she needs, use the following inequality statement:

$p \geq 200$     The points she needs must be greater than or equal to 200.

**Solve each of the following problems using inequalities.**

1. Stella wants to place her money in a high interest money market account. However, she needs at least $1,500 to open an account. Each month, she sets aside some of her earnings in a savings account. In January through June, she added the following amounts to her savings: $145, $203, $210, $120, $102, and $115. Write an inequality which describes the amount of money she can set aside in July to qualify for the money market account.

2. A high school band program will receive $2,000.00 for selling $12,000.00 worth of coupon books. Six band classes participate in the sales drive. Classes 1–5 collect the following amounts of money: $2,400, $2,800, $1,500, $2,320, and $2,550. Write an inequality which describes the amount of money the sixth class must collect so that the band will receive $2,000.

3. A small elevator has a maximum capacity of 1,200 pounds before the cable holding it in place snaps. Six people get on the elevator. Five of their weights follow: 120, 240, 150, 215, and 170. Write an inequality which describes the amount the sixth person can weigh without snapping the cable.

4. A small high school class of 9 students were told they would receive a pizza party if their class average was 90% or higher on the next exam. Students 1–8 scored the following on the exam: 84, 95, 99, 87, 92, 93, 100, and 98. Write an inequality which describes the score the ninth student must make for the class to qualify for the pizza party.

5. Raymond wants to spend his entire credit limit on his credit card. His credit limit is $3,000. He purchases items costing $750, $1,120, $42, $159, $8, and $71. Write an inequality which describes the amounts Raymond can put on his credit card for his next purchases.

# Chapter 4 Review

**Solve each of the following problems.**

1. Deanna is four more than three times older than Ted. The sum of their ages is 60. How old is Ted?

2. The band members sold tickets to their concert performance. Some were $3 tickets, and some were $8 tickets. There were 10 more than twice as many $8 tickets sold as $3 tickets. The total sales were $1,448. How many tickets of each price were sold?

3. Three consecutive integers have a sum of 54. Find the integers.

4. One number is 8 more than the other number. Twice the smaller number is 7 more than the larger number. What are the numbers?

5. The perimeter of a triangle is 48 inches. The second side is four inches longer than the first side. The third side is one inch longer than the second. Find the length of each side.

6. Joe, Craig, and Dylan have a combined weight of 326 pounds. Craig weighs 40 pounds more than Joe. Dylan weighs 12 pounds more than Craig. How many pounds does Craig weigh?

7. Lena and Jodie are sisters, and together they have 56 bottles of nail polish. Lena bought 4 more than half the bottles. How many did Jodie buy?

8. Jim takes great pride in decorating his float for the homecoming parade for his high school. With the $5,000 he has to spend, Jim buys 5,000 carnations at $0.30 each, 4,000 tulips at $0.60 each, and 300 irises at $0.25 each. Write an inequality which describes how many roses, $r$, Jim can buy if roses cost $0.80 each.

9. Mr. Chan wants to sell some or all of his shares of stock in a company. He purchased the 90 shares for $0.50 each last month, and the shares are now worth $3.80 each. Write an inequality which describes how much profit, $p$, Mr. Chan can make by selling his shares.

10. The Jones family traveled 300 miles in 5 hours. What was their average speed?

11. Last year Rikki sang 960 songs with his rock band. How many songs did he sing per month?

12. Erin is looking for a new job. During her interviews, Company A says pay is determined by the equation $y = 13x - 12$, where $x$ is the number of hours worked. How much will Erin make if she can only work ten hours at this company?

13. Cameron lives in Woodstock, Georgia. In his research, he found that the population was 10,050 in the year 2000 and in 2007, the population was 23,000. What is the predicted population for 2014?

14. Timothy bought a car for $5,500 in 1975. In 2000, he learned that, fully restored, his car was worth $65,000. Based on a linear model, how much was Timothy's car worth in 2006? How much will his car be worth in 2009?

**Using dimensional analysis, make the following conversions.**

15. 15 yards to feet

16. 3 pounds to ounces

# Chapter 4 Test

1. Ross is five years older than twice his sister Holly's age. The difference is their ages is 14 years. How old is Holly?

   A 9
   B 23
   C 3
   D 18

2. The sum of two numbers is 27. The larger number is 6 more than twice the smaller number. What are the numbers?

   A 11, 16
   B 19, 8
   C 7, 20
   D 3, 24

3. The perimeter of a rectangle is 292 feet. The length of the rectangle is 4 feet less than 5 times the width. What is the length and width of the rectangle?

   A length = 121, width = 25
   B length = 114.3, width = 23.7
   C length = 25, width = 121
   D length = 121.7, width = 24.3

4. Janet and Artie want to play tug-of-war. Artie pulls with 200 pounds of force while Janet pulls with 60 pounds of force. In order to make this a fair contest, Janet enlists the help of her friends Trudi, Sherri, and Bridget who pull with 20, 25, and 55 pounds respectively. Write an inequality describing the minimum amount Janet's fourth friend, Tommy, must pull to beat Artie.

   A $x > 40$ pounds of force
   B $x < 40$ pounds of force
   C $x > 100$ pounds of force
   D $x < 100$ pounds of force

5. Jesse and Larry entered a pie eating contest. Jesse ate 5 less than twice as many pies as Larry. They ate a total of 16 pies. How many pies did Larry eat?

   A 3.7 pies
   B 9 pies
   C 21 pies
   D 7 pies

6. There is a new bike that Bianca has had her eye on for a few weeks. The bike costs $75. Her allowance is 10 dollars per week. If she saves 60% of her allowance each week, write an inequality that describes the minimum amount of weeks, $y$, that Bianca must save in order to buy that bike.

   A $y > 75 - 0.6\,(10)$
   B $y > 45\,(10)$
   C $y > \dfrac{75}{0.6\,(10)}$
   D $10 > \dfrac{75}{0.6y}$

7. The sum of two numbers is fourteen. The sum of six times the smaller number and two equals four less than the product of three and the larger number. Find the two numbers.

   A 6 and 8
   B 5 and 9
   C 3 and 11
   D 4 and 10

8. Find three consecutive odd numbers whose sum is three hundred and three.

   A 100, 101, 102
   B 99, 101, 103
   C 99, 103, 107
   D 99, 100, 101

9. Tracie and Marcia drove to northern California to see Marcia's sister in Eureka. Tracie drove one hour more than four times as much as Marcia. The trip took a total of 21 driving hours. How many hours did Tracie drive?

**A**  17 hours
**B**  4 hours
**C**  20 hours
**D**  5 hours

10. Alisha climbed a mountain that was 4,760 feet high in 14 hours. What was her average speed per hour?

**A**  476 ft/hr
**B**  4,774 ft/hr
**C**  340 ft/hr
**D**  66,640 ft/hr

11. The new school copy machine makes 3,480 copies per hour. How many copies does this machine make per minute?

**A**  58 copies/minute
**B**  56.8 copies/minute
**C**  208,800 copies/minute
**D**  5 copies/minute

12. Connie drove for 2 hours at a constant speed of 55 mph. How many total miles did she travel?

**A**  27.5 miles
**B**  110 miles
**C**  220 miles
**D**  490 miles

# Chapter 5
# Polynomials

This chapter covers the following IN Algebra I standards:

| Standard 1: | Operations with Real Numbers | A1.1.3 |
|---|---|---|
| Standard 6: | Polynomials | A1.6.1 |
| | | A1.6.2 |
| | | A1.6.3 |
| | | A1.6.4 |
| | | A1.6.5 |

**Polynomials** are algebraic expressions which include **monomials** containing one term, **binomials** which contain two terms, and **trinomials**, which contain three terms. Expressions with more than three terms are called **polynomials.** **Terms** are separated by plus and minus signs.

## EXAMPLES

| Monomials | Binomials | Trinomials | Polynomials |
|---|---|---|---|
| $5f$ | $5t + 20$ | $x^2 + 4x + 3$ | $x^3 - 3x^4 + 3x - 20$ |
| $3x^3$ | $20 - 8g$ | $7x^4 - 6x - 2$ | $p^5 + 4p^3 + p^4 + 20p - 7$ |
| $5g^4$ | $7x^4 + 8x$ | $y^5 + 27y^4 + 200$ | |
| $4$ | $6x^3 - 9x$ | | |

## 5.1 Properties of Addition and Multiplication

The associative, commutative, distributive, identity, and inverse properties of addition and multiplication are listed below by example as a quick refresher.

| Property | Example |
|---|---|
| 1. Associative Property of Addition | $(a + b) + c = a + (b + c)$ |
| 2. Associative Property of Multiplication | $(a \times b) \times c = a \times (b \times c)$ |
| 3. Commutative Property of Addition | $a + b = b + a$ |
| 4. Commutative Property of Multiplication | $a \times b = b \times a$ |
| 5. Distributive Property | $a \times (b + c) = (a \times b) + (a \times c)$ |

**Write the number of the property listed above that describes each of the following statements.**

1. $4 + 5 = 5 + 4$

2. $4 + (2 + 8) = (4 + 2) + 8$

3. $10 (4 + 7) = (10)(4) + (10)(7)$

4. $(2 \times 3) \times 4 = 2 \times (3 \times 4)$

5. $p \times q = q \times p$

6. $x(y + z) = xy + xz$

7. $(m)(n \cdot p) = (m \cdot n)(p)$

8. $z + y = y + z$

## 5.2    Adding and Subtracting Monomials

Two **monomials** are added or subtracted as long as the **variable and its exponent** are the **same**. This is called combining like terms.  Use the same rules you used for adding and subtracting integers

**Example 1:** $\quad 5x + 7x = 12x \quad \begin{array}{r} 3x^5 \\ -9x^5 \\ \hline -6x^5 \end{array} \quad 4x^4 - 20x^4 = -16x^4 \quad \begin{array}{r} 7y \\ +4y \\ \hline 11y \end{array} \quad 6y^3 - 7y^3 = -y^3$

**Remember:**  When the integer in front of the variable is "1", the one is usually not written.  $1x^4$ is the same as $x^4$, and $-1x$ is the same as $-x$.

**Add or subtract the following monomials.**

1. $4x^4 + 7x^4 =$

2. $7t + 9t =$

3. $20y^3 - 4y^3 =$

4. $6g - 9g =$

5. $8y^4 + 9y^4 =$

6. $s^7 + s^7 =$

7. $-4x - 5x =$

8. $5w^4 - w^4 =$

9. $z^5 + 20z^5 =$

10. $-k + 4k =$

11. $3x^4 - 7x^4 =$

12. $20t + 4t =$

13. $-8v^3 + 20v^3 =$

14. $-4x^3 + x^3 =$

15. $20y^5 - 7y^5 =$

16. $\begin{array}{r} y^4 \\ +2y^4 \\ \hline \end{array}$

17. $\begin{array}{r} 4x^3 \\ -9x^3 \\ \hline \end{array}$

18. $\begin{array}{r} 8t^2 \\ +7t^2 \\ \hline \end{array}$

19. $\begin{array}{r} -2y \\ -4y \\ \hline \end{array}$

20. $\begin{array}{r} 5w^2 \\ +8w^2 \\ \hline \end{array}$

21. $\begin{array}{r} 11t^3 \\ -4t^3 \\ \hline \end{array}$

22. $\begin{array}{r} -5z \\ +9z \\ \hline \end{array}$

23. $\begin{array}{r} 4w^5 \\ +w^5 \\ \hline \end{array}$

24. $\begin{array}{r} 7t^3 \\ -6t^3 \\ \hline \end{array}$

25. $\begin{array}{r} 3x \\ +8x \\ \hline \end{array}$

# 5.3   Adding Polynomials

When adding **polynomials,** make sure the exponents and variables are the same on the terms you are combining. The easiest way is to put the terms in columns with **like exponents** under each other. Each column is added as a separate problem. Fill in the blank spots with zeros if it helps you keep the columns straight. You never carry to the next column when adding polynomials.

**Example 2:**   Add $3x^4 + 25$ and $7x^4 + 4x$

$$\begin{array}{r} 3x^4 + 0x + 25 \\ (+)\,7x^4 + 4x + 0 \\ \hline 10x^4 + 4x + 25 \end{array}$$

**Example 3:**   $(5x^3 - 4x) + (-x^3 - 5)$

$$\begin{array}{r} 5x^3 - 4x + 0 \\ (+)\, -x^3 + 0x - 5 \\ \hline 4x^3 - 4x - 5 \end{array}$$

**Add the following polynomials.**

1. $y^4 + 3y + 4$ and $4y^4 + 5$

2. $(7y^4 + 5y - 6) + (4y^4 - 7y + 9)$

3. $-4x^4 + 7x^3 + 5x - 2$ and $3x^4 - x + 4$

4. $-p + 5$ and $7p^4 - 4p + 4$

5. $(w - 4) + (w^4 + 4)$

6. $5t^4 - 7t - 8$ and $9t + 4$

7. $t^5 + t + 9$ and $4t^3 + 5t - 5$

8. $(s^4 + 3s^3 - 4) + (-4s^3 + 5)$

9. $(-v^4 + 8v - 9) + (5v^3 - 6v + 5)$

10. $6m^4 - 4m + 20$ and $m^4 - m - 9$

11. $-x + 5$ and $3x^4 + x - 4$

12. $(9t^4 + 3t) + (-8t^4 - t + 5)$

13. $(3p^5 + 4p^4 - 2) + (-7p^4 - p + 9)$

14. $20s^4 + 24s^3 + 4s$ and $s^4 + s^3 + s$

15. $(-20b^4 + 8b + 4) + (-b^4 + 6b + 20)$

16. $27c^4 - 22c + 7$ and $-8c^4 + 3c - 20$

17. $4c^4 + 7c^3 + 3$ and $5c^4 + 4c^3 + 2$

18. $3x^4 + -25x^3 + 27$ and $8x^3 - 24$

19. $(-x^4 + 4x - 5) + (3x^4 - 3)$

20. $(y^4 - 22y + 20) + (-23y^4 + 7y - 5)$

21. $3d^7 - 5d^3 + 8$ and $4d^5 - 4d^3 - 4$

22. $(6t^7 - t^3 + 28) + (5t^7 + 8t^3)$

23. $5p^4 - 9p + 20$ and $-p^4 - 3p - 7$

24. $40b^3 + 27b$ and $-5b^4 - 7b + 25$

25. $(-4w + 22) + (w^3 + w - 5)$

26. $(47z^4 + 23z + 9) + (z^4 - 4z - 20)$

## 5.4   Subtracting Polynomials

When you subtract polynomials, it is important to remember to change all the signs in the subtracted polynomial (the subtrahend) and then add.

**Example 4:**     $(5y^4 + 9y + 20) - (4y^4 + 6y - 5)$

**Step 1:**     Copy the subtraction problem into vertical form. Make sure you line up the terms with like exponents under each other just like you did for adding polynomials.

$$\begin{array}{r} 5y^4 + 9y + 20 \\ (-)\,4y^4 + 6y - 5 \\ \hline \end{array}$$

**Step 2:**     Change the subtraction sign to addition and all the signs of the subtracted polynomial to the opposite sign. The bottom polynomial in the problem becomes $-4y^4 - 6y + 5$.

**Step 3:**     Add:   $\begin{array}{r} 5y^4 + 9y + 20 \\ (+) - 4y^4 - 6y + 5 \\ \hline y^4 + 3y + 25 \end{array}$

**Subtract the following polynomials.**

1. $(4x^4 + 7x + 4) - (x^4 + 3x + 2)$

2. $(9y - 5) - (5y + 3)$

3. $(-5t^4 + 22t^3 + 3) - (5t^4 - t^3 - 7)$

4. $(-3w^4 + 20w - 7) - (-7w^4 - 7)$

5. $(6a^7 - a^3 + a) - (8a^7 + a^4 - 3a)$

6. $(25c^5 + 40c^4 + 20) - (8c^5 + 7c^4 + 24)$

7. $(7x^4 - 20x) - (-8x^4 + 5x + 9)$

8. $(-9y^4 + 24y^3 - 20) - (3y^3 + y + 20)$

9. $(-3h^4 - 8h + 8) - (7h^4 + 5h + 20)$

10. $(20k^3 - 9) - (k^4 - 5k^3 + 7)$

11. $(x^4 - 7x + 20) - (6x^4 - 7x + 8)$

12. $(24p^4 + 5p) - (20p - 4)$

13. $(-4m - 9) - (6m + 4)$

14. $(4y^4 + 23y^3 - 9y) - (5y^4 + 4y^3 - 8y)$

15. $(8g + 3) - (g^4 + 5g - 7)$

16. $(-9w^3 + 5w) - (-5w^4 - 20w^3 - w)$

17. $(x^4 + 24x^3 - 20) - (4x^4 + 3x^3 + 2)$

18. $(4a^4 + 4a + 4) - (-a^4 + 3a + 3)$

19. $(c + 220) - (3c^4 - 8c + 4)$

20. $(-6v^4 + 24v) - (3v^4 + 4v + 6)$

21. $(3b^4 + 5b^3 + 7) - (8b^3 - 9)$

22. $(7x^4 + 27x^3 - 5) - (-5x^4 + 5x^3)$

23. $(9y^4 - 4) - (22y^4 - 4y - 3)$

24. $(-z^4 - 7z - 9) - (3z^4 - 7z + 7)$

A subtraction of polynomials problem may be stated in sentence form.  Study the examples below.

**Example 5:**    Subtract $-7x^3 + 5x - 3$ from $3x^3 + 5x^4 - 6x$.

**Step 1:**    Copy the problem in columns with terms with the same exponent and variable under each other. Notice the second polynomial in the sentence will be the top polynomial of the problem.

$$3x^3 + 5x^4 - 6x$$
$$(-) - 7x^3 \qquad + 5x - 3$$

Since this is a subtraction problem, change all the signs of the terms in the bottom polynomial. Then add.

$$3x^3 + 5x^4 - 6x$$
$$(+) \, 7x^3 \qquad - 5x + 3$$
$$\overline{10x^3 + 5x^4 - 11x + 3}$$

**Example 6:**    From $6y^4 + 4$ subtract $5y^4 - 3y + 9$

In a problem phrased like this one, the first polynomial will be on top, and the second will be on bottom. Change the signs on the bottom polynomial and then add.

$$6y^4 \qquad + 4$$
$$(-) \, 5y^4 - 3y + 9$$

$$\longrightarrow$$

$$6y^4 \qquad + 4$$
$$(+) - 5y^4 + 3y - 9$$
$$\overline{y^4 + 3y - 5}$$

**Solve the following subtraction problems.**

1. Subtract $3x^4 + 4x - 7$ from $7x^4 + 4$

2. From $7y^3 - 6y + 20$ subtract $9y^3 - 20$

3. From $5m^4 - 5m + 8$ subtract $4m - 3$

4. Subtract $9z^4 + 3z + 4$ from $5z^4 - 8z + 9$

5. Subtract $t^4 + 20t^3 - 7$ from $-t^4 - 4t^3 - 7$

6. Subtract $-8b^3 - 4b + 5$ from $-b^4 + b + 6$

7. From $20y^3 + 40$ subtract $7y^3 - 7$

8. From $25t^4 - 6t - 9$ subtract $5t^4 - 3t + 4$

9. Subtract $3p^4 + p - 4$ from $-8p^4 - 7p + 4$

10. Subtract $x^3 + 9$ from $-4x^4 + 3x^3 + 20$

11. Subtract $24a^4 + 20$ from $-a^4 + a^3 - 2$

12. From $6m^4 + 3m + 2$ subtract $-6m^4 - 3m$

13. From $-3z^4 - 23z^3 - 4$ subtract $-40z^3 + 40$

14. Subtract $20c^4 + 20$ from $9c^4 - 7c + 3$

15. Subtract $b^4 + b - 7$ from $7b^4 - 5b + 7$

16. Subtract $-3x - 5$ from $3x^4 + x + 20$

17. From $27y^4 + 4$ subtract $5y^4 + 3y + 8$

18. Subtract $3g^4 - 7g + 7$ from $20g^4 - 3g - 5$

19. From $-8m^4 - 9m$ subtract $3m^4 + 8$

20. Subtract $x + 2$ from $7x + 7$

21. Subtract $c^4 + c + 4$ from $-c^4 - c - 4$

22. From $6t^4 + 9t^3 - 5t + 4$ subtract $t^3 + 3t$

## 5.5   Multiplying Monomials

When two monomials have the **same variable**, you can multiply them. Then, add the **exponents** together. If the variable has no exponent, it is understood that the exponent is 1.

**Example 7:**      $5x^5 \times 3x^4 = 15x^9$                    $4y \times 7y^4 = 28y^5$

**Multiply the following monomials.**

1.  $6a \times 20a^7$

2.  $4x^6 \times 7x^3$

3.  $5y^3 \times 3y^4$

4.  $20t^4 \times 4t^4$

5.  $4p^7 \times 5p^4$

6.  $20b^4 \times 9b$

7.  $3c^3 \times 3c^3$

8.  $4d^9 \times 20d^4$

9.  $6k^3 \times 7k^4$

10.  $8m^7 \times m$

11.  $22z \times 4z^8$

12.  $3w^5 \times 6w^7$

13.  $5x^5 \times 7x^3$

14.  $7n^4 \times 3n^3$

15.  $9w^8 \times w$

16.  $20s^6 \times 7s^3$

17.  $5d^7 \times 5d^7$

18.  $7y^4 \times 9y^6$

19.  $8t^{20} \times 3t^7$

20.  $6p^9 \times 4p^3$

21.  $x^3 \times 4x^3$

**When problems include negative signs, follow the rules for multiplying integers.**

22. $-8s^5 \times 7s^3$

23. $-6a \times -20a^7$

24. $5x \times -x$

25. $-3y^4 \times -y^3$

26. $-7b^4 \times 3b^7$

27. $20c^5 \times -4c$

28. $-5t^3 \times 9t^3$

29. $20d \times -9d^8$

30. $-3g^6 \times -4g^3$

31. $-8s^5 \times 8s^3$

32. $-d^3 \times -4d$

33. $22p \times -4p^7$

34. $-7x^8 \times -3x^3$

35. $9z^5 \times 8z^5$

36. $-5w \times -7w^9$

37. $-7y^5 \times 6y^4$

38. $20x^3 \times -8x^7$

39. $-a^5 \times -a$

40. $-8k^4 \times 3k$

41. $-27t^4 \times -t^5$

42. $3x^9 \times 20x^4$

## 5.6   Multiplying Monomials with Different Variables

**Warning: You cannot add the exponents of variables that are different.**

**Example 8:**                    $(-5wx)(6w^3x^4)$

To work this problem, first multiply the whole numbers: $-5 \times 6 = -30$. Then multiply the $w$'s: $w \times w^3 = w^4$. Last, multiply the $x$'s: $x \times x^4 = x^5$. The answer is $-30w^4x^5$.

**Multiply the following monomials.**

1. $(4x^4y^4)(-5xy^3) =$

2. $(20p^3q^5)(4p^4q) =$

3. $(-3t^5v^4)(t^4v) =$

4. $(8w^3z^4)(3wz) =$

5. $(-4st^6)(-9s^4t) =$

6. $(xy^3)(5x^4y^4) =$

7. $(7y^4z)(3y^5z^4) =$

8. $(-3a^4b^4)(-5ab^3) =$

9. $(-7c^3d^4)(4c^5d^7) =$

10. $(20x^5y^4)(3x^3y) =$

11. $(6f^3g^7)(-f^3g) =$

12. $(-5a^3v^5)(9a^5v) =$

13. $(7m^9n^7)(8m^4n^5) =$

14. $(8w^7y^3)(3wy) =$

15. $(4x^5z^4)(-20x^4z^5) =$

16. $(-5a^8c^{20})(4a^4c) =$

17. $(-bd^6)(-b^4d) =$

18. $(3x^5y^4)(20x^3y^3) =$

19. $(20p^4y)(7p^7y^3) =$

20. $(-4a^8x^4)(6ax^4) =$

21. $(9c^5d^3)(-4c^4d^4) =$

**Multiplying three monomials works the same way. The first one is done for you.**

22. $(3st)(5s^3t^4)(4s^4t^5) = 60s^8t^{10}$

23. $(xy)(x^4y^4)(4x^3y^4) =$

24. $(4a^4b^4)(a^3b^3)(4ab) =$

25. $(5y^4z^5)(4y^3)(4z^4) =$

26. $(7cd^3)(3c^4d^4)(d^4) =$

27. $(4w^4x^3)(3x^5)(4w^3) =$

28. $(a^5d^4)(ad)(a^4d^3) =$

29. $(8x^3t)(4t^5)(x^4t^4) =$

30. $(p^4y^4)(5py)(p^3y^3) =$

31. $(5x^3y)(7xy^3)(4y^5) =$

32. $(9xy^4)(x^4y^3)(4x^4y) =$

33. $(6p^3t)(4t^3)(p^4) =$

34. $(3bc)(b^4c)(5c^3) =$

35. $(4y^5z^7)(4y^6)(y^4z^4) =$

36. $(5p^3r^3)(5r^4)(p^4r) =$

37. $(a^5z^6)(6a^4z^4)(3z^3) =$

38. $(7c^3)(6d^4)(4c^4d) =$

39. $(20s^8t^4)(3st)(s^4t^3) =$

40. $(3a^3b^5)(4b^3)(3a^4) =$

41. $(7wz)(w^3z^3)(3w^4z^3) =$

## 5.7   Dividing Monomials

When simplifying monomial fractions with exponents, all exponents need to be positive. If there are negative exponents in your answer, put the base with its negative exponent below the fraction line and remove the negative sign. Two variables that are alike should not appear in both the denominator and numerator of a simplified expression. If you have the same variable in both the denominator and numerator of the fraction, the expression is not in simplest terms.

**Example 9:** $\dfrac{55x^4y^4}{11x^6y^6}$

**Step 1:** Reduce the whole numbers first. $\dfrac{55}{11} = 5$

**Step 2:** Simplify the $x$'s. $\dfrac{x^4}{x^6} = x^{4-6} = x^{-2} = \dfrac{1}{x^2}$

**Step 3:** Simplify the $y$'s. $\dfrac{y^4}{y^6} = y^{4-6} = y^{-2} = \dfrac{1}{y^2}$

Therefore $\dfrac{55x^4y^4}{11x^6y^6} = \dfrac{5}{x^2y^2}$

**Simplify the expressions below. All answers should only have positive exponents.**

1. $\dfrac{7xy^3}{x(4x^3)y^5}$

2. $\dfrac{4a^4b^7}{3a^5b^4}$

3. $\dfrac{8(4a^4)b^5}{9ab^6}$

4. $\dfrac{26(x^4y^5)^3}{40xy}$

5. $\dfrac{20a^5b^4}{7a^6b^3}$

6. $\dfrac{7(9x^4y^3)}{5(x^4y^4)^4}$

7. $\dfrac{24(3a^4)b^4}{6a^4b^4}$

8. $\dfrac{(6x^3y^5)^4}{(4x^7y)^3}$

9. $\dfrac{27a^4b^3}{3a^7b^6}$

10. $\dfrac{33x^7y^3}{44x^8y^7}$

11. $\dfrac{27(4a^6b^8)}{42a^3b^6}$

12. $\dfrac{30x^5y^4}{6(x^4y^4)^4}$

13. $\dfrac{20(9ab^7)}{40a^4b^4}$

14. $\dfrac{20x^{20}y^8}{57x^7y^3}$

15. $\dfrac{(a^5b^8)^5}{a^9b^8}$

16. $\dfrac{8(x^3y^7)}{9x^4y^5}$

17. $\dfrac{20(a^3b^5)}{5a^7b^4}$

18. $\dfrac{48(x^4y^7)^4}{42x^3y^5}$

## 5.8 Extracting Monomial Roots

When finding the roots of monomial expressions, you must first divide the monomial expression into separate parts. Then, simplify each part of the expression.

**Note: To find the square root of any variable raised to a positive exponent, simply divide the exponent by 2. For example, $\sqrt{y^{10}} = y^5$.**

**Example 10:** $\sqrt{25x^6 y^4 z^6}$

**Step 1:** Break each component apart. $\left(\sqrt{25}\right)\left(\sqrt{x^6}\right)\left(\sqrt{y^4}\right)\left(\sqrt{z^6}\right)$

**Step 2:** Solve for each component. $\left(\sqrt{25} = 5\right)\left(\sqrt{x^6} = x^3\right)\left(\sqrt{y^4} = y^2\right)\left(\sqrt{z^6} = z^3\right)$

**Step 3:** Recombine the simplified expressions. $(5)\left(x^3\right)\left(y^2\right)\left(z^3\right) = 5x^3 y^2 z^3$

**Simplify the problems below.**

1. $\sqrt{4a^2 b^4 c^8}$

2. $\sqrt{49h^{24} i^6 j^4}$

3. $\sqrt{121p^{20} q^{24} r^6}$

4. $\sqrt{36a^{28} b^{10} c^6}$

5. $\sqrt{144t^{44} u^{30} v^2}$

6. $\sqrt{36k^6 l^{26} m^{20}}$

7. $\sqrt{25s^6 t^{24} u^{32}}$

8. $\sqrt{81x^6 y^{20} z^{44}}$

9. $\sqrt{49a^6 b^4 c^8}$

10. $\sqrt{169u^{12} v^{24} w^{28}}$

11. $\sqrt{64x^{44} y^{20} z^6}$

12. $\sqrt{4d^{30} e^{54} f^{10}}$

13. $\sqrt{f^4 g^6 h^{26}}$

14. $\sqrt{900l^{50} m^{26} n^4}$

15. $\sqrt{400g^{40} h^{26} i^{36}}$

16. $\sqrt{25a^{54} b^6 c^{46}}$

17. $\sqrt{16j^{24} k^8 l^{20}}$

18. $\sqrt{9q^4 r^{20} s^{34}}$

## 5.9    Monomial Roots with Remainders

Monomial roots which are not easily simplified under the square root symbol will also sometimes be encountered. Powers may be raised to odd numbers. In addition, the coefficients may not be perfect squares. Follow the example below to understand how to simplify these types of problems.

**Example 11:**    Simplify $\sqrt{40x^7y^{11}z^{23}}$

   **Step 1:**    Begin by simplifying the coefficient. $\sqrt{40} = \left(\sqrt{4}\right)\left(\sqrt{10}\right)$, $\sqrt{4} = 2$, so
   $\sqrt{40} = 2\sqrt{10}$

   **Step 2:**    Simplify the variable with exponents.

$$\sqrt{x^7} = \left(\sqrt{x^6}\right)\left(\sqrt{x}\right), \sqrt{x^6} = x^3, \text{ so } \sqrt{x^7} = x^3\sqrt{x}$$

$$\sqrt{y^{11}} = \left(\sqrt{y^{10}}\right)\left(\sqrt{y}\right), \sqrt{y^{10}} = y^5, \text{ so } \sqrt{y^{11}} = y^5\sqrt{y}$$

$$\sqrt{z^{23}} = \left(\sqrt{z^{22}}\right)\left(\sqrt{z}\right), \sqrt{z^{22}} = z^{11}, \text{ so } \sqrt{z^{23}} = z^{11}\sqrt{z}$$

   **Step 3:**    Recombine the simplified expressions. $2x^3y^5z^{11}\sqrt{10xyz}$

**Simplify the following square root expressions.**

1. $\sqrt{57d^{25}e^{27}f^{22}}$

2. $\sqrt{140h^{26}i^{20}j^9}$

3. $\sqrt{27x^{44}y^{42}z^9}$

4. $\sqrt{75p^{22}q^8r^{21}}$

5. $\sqrt{48k^{47}l^{27}m^3}$

6. $\sqrt{75s^{23}t^7u^{28}}$

7. $\sqrt{63a^8b^{27}c^{42}}$

8. $\sqrt{20p^3q^{44}r^{29}}$

9. $\sqrt{80a^{220}b^{20}c^{27}}$

10. $\sqrt{64m^8n^3p^{22}}$

11. $\sqrt{88r^{27}s^{22}t^{23}}$

12. $\sqrt{40g^{42}h^{25}j^{28}}$

13. $\sqrt{90v^3w^{20}x^{24}}$

14. $\sqrt{50d^7e^9f^{23}}$

15. $\sqrt{45x^{28}y^6z^{23}}$

16. $\sqrt{32a^6b^{23}c^7}$

17. $\sqrt{74j^{24}k^{27}m^7}$

18. $\sqrt{20q^{24}r^{27}s^7}$

## 5.10   Multiplying Monomials by Polynomials

In the chapter on solving multi-step equations, you learned to remove parentheses by multiplying the number outside the parentheses by each term inside the parentheses: $4(5x - 8) = 9x - 25$. Multiplying monomials by polynomials works the same way.

**Example 12:**   $-7t(4t^4 - 8t + 20)$

**Step 1:**   Multiply $-7t \times 4t^4 = -28t^5$

**Step 2:**   Multiply $-7t \times -8t = 56t^2$

**Step 3:**   Multiply $-7t \times 20 = -140t$

**Step 4:**   Arrange the answers horizontally in order: $-28t^5 + 56t^2 - 140t$

**Remove parentheses in the following problems.**

1. $3x(3x^4 + 5x - 2)$

2. $5y(y^3 - 8)$

3. $8a^4(4a^4 + 3a + 4)$

4. $-7d^3(d^4 - 7d)$

5. $4w(-5w^4 + 3w - 9)$

6. $9p(p^3 - 6p + 7)$

7. $-20b^4(-4b + 7)$

8. $4t(t^4 - 5t - 20)$

9. $20c(5c^4 + 3c - 8)$

10. $6z(4z^5 - 7z^4 - 5)$

11. $-20t^4(3t^4 + 7t + 6)$

12. $c(-3c - 7)$

13. $3p(-p^4 + p^3 - 20)$

14. $-k^4(4k + 5)$

15. $-3(5m^4 - 7m + 9)$

16. $6x(-8x^3 + 20)$

17. $-w(w^4 - 5w + 8)$

18. $4y(7y^4 - y)$

19. $3d(d^7 - 8d^3 + 5)$

20. $-7t(-5t^4 - 9t + 2)$

21. $8(4w^4 - 20w + 5)$

22. $3y^4(y^4 - 22)$

23. $v^4(v^4 + 3v + 3)$

24. $9x(4x^3 + 3x + 2)$

25. $-7d(5d^4 + 8d - 4)$

26. $-k^4(-3k + 6)$

27. $3x(-x^4 - 7x + 7)$

28. $5z(5z^5 - z - 8)$

29. $-7y(20y^3 - 3)$

30. $4b^4(8b^4 + 5b + 5)$

## 5.11    Dividing Polynomials by Monomials

**Example 13:**    $\dfrac{-8wx + 6x^2 - 16wx^2}{2wx}$

    **Step 1:**    Rewrite the problem. Divide each term from the top by the denominator, $2wx$.

$$\frac{-8wx}{2wx} + \frac{6x^2}{2wx} + \frac{-16wx^2}{2wx}$$

    **Step 2:**    Simplify each term in the problem. Then combine like terms.

$$-4 + \frac{3x}{w} - 8x$$

**Simplify each of the following.**

1. $\dfrac{bc^4 - 9bc - 4b^4c^4}{4bc}$

2. $\dfrac{3jk^4 + 24k + 20j^4k}{3jk}$

3. $\dfrac{7x^4y - 9xy^4 + 4y^3}{4xy}$

4. $\dfrac{26st^4 + st - 24s}{5st}$

5. $\dfrac{5wx^4 + 6wx - 24w^3}{4wx}$

6. $\dfrac{cd^4 + 20cd^3 + 26c^4}{4cd}$

7. $\dfrac{y^4z^3 - 4yz - 9z^4}{-4yz^4}$

8. $\dfrac{a^4b + 4ab^4 - 25ab^3}{4a^4}$

9. $\dfrac{pr^4 + 6pr + 9p^4r^4}{4pr^4}$

10. $\dfrac{6xy^4 - 3xy + 29x^4}{-3xy}$

11. $\dfrac{6x^4y + 24xy - 45y^4}{6xy}$

12. $\dfrac{7m^4n - 20mn - 47n^4}{7mn}$

13. $\dfrac{st^4 - 20st - 26s^4t^4}{4st}$

14. $\dfrac{8jk^4 - 25jk - 63j^4}{8jk}$

## 5.12   Removing Parentheses and Simplifying

In the following problem, you must multiply each term inside the parentheses by the numbers and variables outside the parentheses, and then add the polynomials to simplify the expressions.

**Example 14:**   $9x\left(4x^4 - 7x + 8\right) - 3x\left(5x^4 + 3x - 9\right)$

**Step 1:**   Multiply to remove the first set of parentheses.

$9x\left(4x^4 - 7x + 8\right) = 36x^5 - 63x^2 + 72x$

**Step 2:**   Multiply to remove the second set of parentheses.

$-3x\left(5x^4 + 3x - 9\right) = -15x^5 - 9x^2 + 27x$

**Step 3:**   Copy each polynomial in columns, making sure the terms with the same variable and exponent are under each other. Add to simplify.

$$\begin{array}{r} 36x^5 - 63x^2 + 72x \\ (+) - 15x^5 - 9x^2 + 27x \\ \hline 21x^5 - 72x^2 + 99x \end{array}$$

**Remove the parentheses and simplify the following problems.**

1.   $5t\left(t + 8\right) + 7t\left(4t^4 - 5t + 2\right)$

2.   $-7y\left(3y^4 - 7y + 3\right) - 6y\left(y^4 - 5y - 5\right)$

3.   $-3\left(3x^4 + 5x\right) + 7x\left(x^4 + 3x + 4\right)$

4.   $4b\left(7b^4 - 9b - 2\right) - 3b\left(5b + 3\right)$

5.   $9d^4\left(3d + 5\right) - 8d\left(3d^4 + 5d + 7\right)$

6.   $7a\left(3a^4 + 3a + 2\right) - \left(-4a^4 + 7a - 5\right)$

7.   $3m\left(m + 8\right) + 9\left(5m^4 + m + 5\right)$

8.   $5c^4\left(-6c^4 - 3c + 4\right) - 8c\left(7c^3 + 4c\right)$

9.   $-9w\left(-w + 2\right) - 5w\left(3w - 7\right)$

10.   $6p\left(4p^4 - 5p - 6\right) + 3p\left(p^4 + 6p + 20\right)$

## 5.13   Multiplying Two Binomials

When you multiply two binomials such as $(x + 6)(x - 7)$, you must multiply each term in the first binomial by each term in the second binomial. The easiest way is to use the **FOIL** method. If you can remember the word **FOIL**, it can help you keep order when you multiply. The "**F**" stands for **first**, "**O**" stands for **outside**, "**I**" stands for **inside**, and "**L**" stands for **last**.

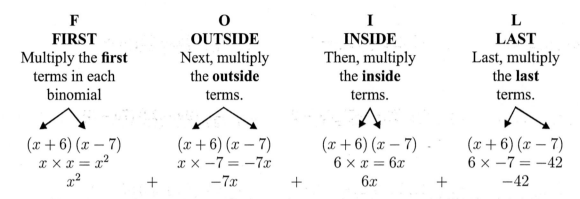

| F | O | I | L |
|---|---|---|---|
| **FIRST** | **OUTSIDE** | **INSIDE** | **LAST** |
| Multiply the **first** terms in each binomial | Next, multiply the **outside** terms. | Then, multiply the **inside** terms. | Last, multiply the **last** terms. |

$$(x + 6)(x - 7) \qquad (x + 6)(x - 7) \qquad (x + 6)(x - 7) \qquad (x + 6)(x - 7)$$
$$x \times x = x^2 \qquad x \times -7 = -7x \qquad 6 \times x = 6x \qquad 6 \times -7 = -42$$
$$x^2 \quad + \quad -7x \quad + \quad 6x \quad + \quad -42$$

Now just combine like terms, $6x - 7x = -x$, and write your answer.
$(x + 6)(x - 7) = x^2 - x - 42$.

**Note:** It is customary for mathematicians to write polynomials in descending order. That means that the term with the highest-number exponent comes first in a polynomial. The next highest exponent is second and so on. When you use the **FOIL** method, the terms will always be in the customary order. You just need to combine like terms and write your answer.

**Multiply the following binomials.**

1. $(y - 8)(y + 3)$

2. $(4x + 5)(x + 20)$

3. $(5b - 3)(3b - 5)$

4. $(6g + 4)(g - 20)$

5. $(8k - 7)(-5k - 3)$

6. $(9v - 4)(3v + 5)$

7. $(20p + 4)(5p + 3)$

8. $(3h - 20)(-4h - 7)$

9. $(w - 5)(w - 8)$

10. $(6x + 2)(x - 4)$

11. $(7t + 3)(4t - 2)$

12. $(5y - 20)(5y + 20)$

13. $(a + 6)(3a + 7)$

14. $(3z - 9)(z - 5)$

15. $(7c + 4)(6c + 7)$

16. $(y + 3)(y - 3)$

17. $(4w - 7)(5w + 6)$

18. $(8x + 2)(x - 5)$

19. $(6t - 20)(5t - 5)$

20. $(7b + 6)(6b + 4)$

21. $(4z + 2)(20z + 5)$

22. $(22w - 9)(w + 3)$

23. $(7d - 20)(20d + 20)$

24. $(20g + 4)(g - 4)$

25. $(5p + 8)(4p + 3)$

26. $(m + 7)(m - 7)$

27. $(9b - 9)(4b - 2)$

28. $(z + 3)(3z + 7)$

29. $(8y - 7)(y - 3)$

30. $(20x + 7)(3x - 2)$

31. $(3t + 2)(t + 20)$

32. $(4w - 20)(9w + 8)$

33. $(9s - 4)(s + 5)$

34. $(5k - 2)(9k + 20)$

35. $(h + 24)(h - 4)$

36. $(3x + 8)(8x + 3)$

37. $(4v - 6)(4v + 6)$

38. $(4x + 9)(4x - 3)$

39. $(k - 2)(6k + 24)$

40. $(3w + 22)(4w + 4)$

41. $(9y - 20)(7y - 3)$

42. $(6d + 23)(d - 2)$

43. $(8h + 3)(4h + 5)$

44. $(7n + 20)(7n - 7)$

45. $(6z + 7)(z - 9)$

46. $(5p + 7)(4p - 20)$

47. $(b + 4)(7b + 8)$

48. $(20y - 3)(9y - 8)$

# Chapter 5 Review

## Simplify.

1. $3a^4 + 20a^4$

2. $(8x^4y^5)(20xy^7)$

3. $-6z^4(z+3)$

4. $(5b^4)(7b^3)$

5. $8x^4 - 20x^4$

6. $(7p-5)-(3p+4)$

7. $(3w^3y^4)(5wy^7)$

8. $25d^5 - 20d^5$

9. $(8w-5)(w-9)$

10. $27t^4 + 5t^4$

11. $(8c^5)(20c^4)$

12. $(20x+4)(x+7)$

13. $5y(5y^4 - 20y + 4)$

14. $(9a^5b)(4ab^3)(ab)$

15. $(7w^6)(20w^{20})$

16. $9x^3 + 24x^3$

17. $27p^7 - 22p^7$

18. $(3s^5t^4)(5st^3)$

19. $(5d+20)(4d+8)$

20. $5w(-3w^4 + 8w - 7)$

21. $45z^6 - 20z^6$

22. $-8y^3 - 9y^3$

23. $(8x^5)(8x^7)$

24. $28p^4 + 20p^4$

25. $(a^4v)(4av)(a^3v^6)$

26. $(3c^4)(6c^9)$

27. $(5x^7y^3)(4xy^3)$

28. Add $4x^4 + 20x$ and $7x^4 - 9x + 4$

29. $5t(6t^4 + 5t - 6) + 9t(3t+3)$

30. Subtract $y^4 + 5y - 6$ from $3y^4 + 8$

31. $4x(5x^4 + 6x - 3) + 5x(x+3)$

32. $(6t-5)-(6t^4 + t - 4)$

33. $(5x+6)+(8x^4 - 4x + 3)$

34. Subtract $7a - 4$ from $a + 20$

35. $(-4y+5)+(5y-6)$

36. $4t(t+6) - 7t(4t+8)$

37. Add $3c - 5$ and $c^4 - 3c - 4$

38. $4b(b-5)-(b^4 + 4b + 2)$

39. $(6k^4 + 7k)+(k^4 + k + 20)$

40. $(q^4r^3)(3qr^4)(4q^5r)$

41. $(7df)(d^5f^4)(4df)$

42. $(8g^4h^3)(g^3h^6)(6gh^3)$

43. $(9v^4x^3)(3v^6x^4)(4v^5x^5)$

44. $(3n^4m^4)(20n^4m)(n^3m^8)$

45. $(22t^4a^4)(5t^3a^9)(4t^6a)$

46. $\dfrac{24(4a^3)b}{3a^4b^{-4}}$

47. $\dfrac{8(g^3h^3)}{5(g^4h)^{-4}}$

48. $\dfrac{26(m^4n^3)^4}{5(m^4n)^{-4}}$

49. $\dfrac{25p^3q^3}{4p^4q}$

50. $\dfrac{9(e^5h^{-4})^{-4}}{36e^4h^7}$

51. $\dfrac{44x^3y^5}{154(x^{-3}y^8)^4}$

## Identify the property used in each equation below.

52. $x(y+z) = xy + xz$

53. $(m)(n \cdot p) = (m \cdot n)(p)$

54. $a \times b = b \times a$

# Chapter 5 Test

1. $2x^2 + 5x^2 =$

   **A** $10x^4$
   **B** $7x^4$
   **C** $7x^2$
   **D** $10x^2$

2. $-8m^3 + m^3 =$

   **A** $-8m^6$
   **B** $-8m^9$
   **C** $-9m^6$
   **D** $-7m^3$

3. $(6x^3 + x^2 - 5) + (-3x^3 - 2x^2 + 4) =$

   **A** $3x^3 - x^2 - 1$
   **B** $3x^3 - 3x^2 - 1$
   **C** $3x^3 - 3x^2 - 9$
   **D** $-3x^3 - 3x^2 - 1$

4. $(-7c^2 + 5c + 3) + (-c^2 - 7c + 2) =$

   **A** $-3x^3 - 3x^2 - 1$
   **B** $-8c^2 - 2c + 5$
   **C** $-6c^2 - 12c + 5$
   **D** $-8c^2 - 12c + 5$

5. $(5x^3 - 4x^2 + 5) - (-2x^3 - 3x^2) =$

   **A** $3x^3 + x^2 + 5$
   **B** $3x^3 - 7x^2 + 5$
   **C** $7x^3 - x^2 + 5$
   **D** $7x^3 - 7x^2 + 5$

6. $(-z^3 - 4z^2 - 6) - (3z^3 - 6z + 5) =$

   **A** $-4z^3 - 4z^2 + 6z - 11$
   **B** $-2z^3 - 10z - 1$
   **C** $-4z^3 - 10z^2 - 1$
   **D** $-2z^2 + 2z - 11$

7. $(-7d^5)(-3d^2) =$

   **A** $-21d^7$
   **B** $21d^{10}$
   **C** $21d^7$
   **D** $-21d^{10}$

8. $(-5c^3 d)(3c^5 d^3)(2cd^4) =$

   **A** $30c^{15}d^8$
   **B** $15c^8 d^{12}$
   **C** $-17c^{15}d^{12}$
   **D** $-30c^9 d^8$

9. $-11j^2 \times -j^4 =$

   **A** $11j^6$
   **B** $11j^8$
   **C** $-11j^6$
   **D** $-11j^8$

10. $-6m^2(7m^2 + 5m - 6) =$

   **A** $-42m^2 + 30m^3 - 36$
   **B** $-42m^4 - 30m^3 + 36m^2$
   **C** $-13m^4 - m^2 + 36m^2$
   **D** $42m^4 - 30m^3 - 36m^2$

11. $-h^2(-4h + 5) =$

   **A** $-4h^3 - 5h^2$
   **B** $4h^3 - 5h^2$
   **C** $-5h^2 - 5h^2$
   **D** $-5h^3 - 5h^2$

12. $\dfrac{4xy^2 - 6xy + 8x^2 y}{2xy} =$

   **A** $2xy - 3 + 4x$
   **B** $2y - 3 + 4xy$
   **C** $2y - 3 + 4x$
   **D** $2xy - 3 + 4x^2$

13. $\dfrac{3cd^3 + 6c^2d - 12cd}{3cd} =$

    **A**  $cd + 3c - 4cd$
    **B**  $d^2 + 2c - 4cd$
    **C**  $cd^2 + 2c - 4cd$
    **D**  $d^2 + 2c - 4$

14. $4m(m - 5) + 3m(2m^2 - 6m + 4) =$

    **A**  $6m^3 - 14m^2 - 8m$
    **B**  $-8m^2 - 8m - 1$
    **C**  $7m - 14m^2 - 1$
    **D**  $10m^2 - 26m - 20$

15. $2h(3h^2 - 5h - 2) + 4h(h^2 + 6h + 8) =$

    **A**  $6h^3 + 19h^2 + 28h$
    **B**  $-8m^2 - 8m - 1$
    **C**  $7m - 14m^2 - 1$
    **D**  $10h^3 + 14h^2 + 28h$

16. Multiply the following binomials and simplify.

    $(x - 3)(x + 3)$

    **A**  $x^2 - 3x + 3x - 9$
    **B**  $x^2 - 9$
    **C**  $x^2 + 9$
    **D**  $x^2 + 6x + 9$

17. Multiply the following binomials and simplify.

    $(x + 9)(x + 1)$

    **A**  $x^2 + 10x + 9$
    **B**  $x^2 + 10x + 10$
    **C**  $x^2 + 9x + 9$
    **D**  $x^2 + 9x + x + 9$

18. Which property is demonstrated by the expression below.
    $2 + (3 + 8) = (2 + 3) + 8$

    **A**  commutative property of addition
    **B**  associative property of addition
    **C**  distributive property
    **D**  identity property of addition

19. Is the expression $15 \times 8$ equivalent to the expression $8 \times 15$?

    **A**  Yes because of the commutative property.
    **B**  Yes because of the associative property.
    **C**  Yes because of the distributive property.
    **D**  Yes because of the inverse property.

20. Is the equation $5(x - 6) = -2$ equivalent to the equation $5x - 30 = -2$?

    **A**  Yes because of the commutative property.
    **B**  Yes because of the distributive property.
    **C**  Yes because of the associative property.
    **D**  Yes because of the inverse property.

# Chapter 6
# Factoring

This chapter covers the following IN Algebra I standards:

| Standard 6: | Polynomials | A1.6.6 |
| | | A1.6.7 |
| Standard 7: | Algebraic Fractions | A1.7.1 |
| | | A1.7.2 |

## 6.1 Finding the Greatest Common Factor of Polynomials

In a multiplication problem, the numbers multiplied together are called **factors**. The answer to a multiplication problem is a called the **product**.

In the multiplication problem $5 \times 4 = 20$, 5 and 4 are factors and 20 is the product.

If we reverse the problem, $20 = 5 \times 4$, we say we have **factored** 20 into $5 \times 4$.

In this chapter, we will factor **polynomials**.

**Example 1:** Find the greatest common factor of $2y^3 + 6y^2$.

**Step 1:** Look at the whole numbers. The greatest common factor of 2 and 6 is 2. Factor the 2 out of each term.

$$2\left(y^3 + 3y^2\right)$$

**Step 2:** Look at the remaining terms, $y^3 + 3y^2$. What are the common factors of each term?

$$\begin{array}{rcl} y^3 &=& y \times \boxed{y \times y} \\ 3y^2 &=& 3 \times \boxed{y \times y} \end{array} \longleftarrow \text{common factors} = y^2$$

**Step 3:** Factor 2 and $y^2$ out of each term: $2y^2(y + 3)$

**Check:** $2y^2(y + 3) = 2y^3 + 6y^2$

**Factor by finding the greatest common factor in each of the following.**

1. $6x^4 + 18x^2$

2. $14y^3 + 7y$

3. $4b^5 + 12b^3$

4. $10a^3 + 5$

5. $2y^3 + 8y^2$

6. $6x^4 - 12x^2$

7. $18y^2 - 12y$

8. $15a^3 - 25a^2$

9. $4x^3 + 16x^2$

10. $6b^2 + 21b^5$

11. $27m^3 + 18m^4$

12. $100x^4 - 25x^3$

13. $4b^4 - 12b^3$

14. $18c^2 + 24c$

15. $20y^3 + 30y^5$

16. $16x^2 - 24x^5$

17. $15a^4 - 25a^2$

18. $24b^3 + 16b^6$

19. $36y^4 + 9y^2$

20. $42x^3 + 49x$

Factoring larger polynomials with 3 or 4 terms works the same way.

**Example 2:**   $4x^5 + 16x^4 + 12x^3 + 8x^2$

**Step 1:**   Find the greatest common factor of the whole numbers. 4 can be divided evenly into 4, 16, 12, and 8; therefore, 4 is the greatest common factor.

**Step 2:**   Find the greatest common factor of the variables. $x^5$, $x^4$, $x^3$, and $x^2$ can be divided by $x^2$, the lowest power of $x$ in each term.

$$4x^5 + 16x^4 + 12x^3 + 8x^2 = 4x^2 \left( x^3 + 4x^2 + 3x + 2 \right)$$

**Factor each of the following polynomials.**

1. $5a^3 + 15a^2 + 20a$

2. $18y^4 + 6y^3 + 24y^2$

3. $12x^5 + 21x^3 + x^2$

4. $6b^4 + 3b^3 + 15b^2$

5. $14c^3 + 28c^2 + 7c$

6. $15b^4 - 5b^2 + 20b$

7. $t^3 + 3t^2 - 5t$

8. $8a^3 - 4a^2 + 12a$

9. $16b^5 - 12b^4 - 10b^2$

10. $20x^4 + 16x^3 - 24x^2 + 28x$

11. $40b^7 + 30b^5 - 50b^3$

12. $20y^4 - 15y^3 + 30y^2$

13. $4m^5 + 8m^4 + 12m^3 + 6m^2$

14. $16x^5 + 20x^4 - 12x^3 + 24x^2$

15. $18y^4 + 21y^3 - 9y^2$

16. $3n^5 + 9n^3 + 12n^2 + 15n$

17. $4d^6 - 8d^2 + 2d$

18. $10w^2 + 4w + 2$

19. $6t^3 - 3t^2 + 9t$

20. $25p^5 - 10p^3 - 5p^2$

21. $18x^4 + 9x^2 - 36x$

22. $6b^4 - 12b^2 - 6b$

23. $y^3 + 3y^2 - 9y$

24. $10x^5 - 2x^4 + 4x^2$

**Example 3:**    Find the greatest common factor of $4a^3b^2 - 6a^2b^2 + 2a^4b^3$

**Step 1:**    The greatest common factor of the whole numbers is 2.

$$4a^3b^2 - 6a^2b^2 + 2a^4b^3 = 2\left(2a^3b^2 - 3a^2b^2 + a^4b^3\right)$$

**Step 2:**    Find the lowest power of each variable that is in each term. Factor them out of each term. The lowest power of $a$ is $a^2$. The lowest power of $b$ is $b^2$.

$$4a^3b^2 - 6a^2b^2 + 2a^4b^3 = 2a^2b^2\left(2a - 3 + a^2b\right)$$

## Factor each of the following polynomials.

1. $3a^2b^2 - 6a^3b^4 + 9a^2b^3$

2. $12x^4y^3 + 18x^3y^4 - 24x^3y^3$

3. $20x^2y - 25x^3y^3$

4. $12x^2y - 20x^2y^2 + 16xy^2$

5. $8a^3b + 12a^2b + 20a^2b^3$

6. $36c^4 + 42c^3 + 24c^2 - 18c$

7. $14m^3n^4 - 28m^3n^2 + 42m^2n^3$

8. $16x^4y^2 - 24x^3y^2 + 12x^2y^2 - 8xy^2$

9. $32c^3d^4 - 56c^2d^3 + 64c^3d^2$

10. $21a^4b^3 + 27a^2b^3 + 15a^3b^2$

11. $4w^3t^2 + 6w^2t - 8wt^2$

12. $5pw^3 - 2p^2q^2 - 9p^3q$

13. $49x^3t^3 + 7xt^2 - 14xt^3$

14. $9cd^4 - 3d^4 - 6c^2d^3$

15. $12a^2b^3 - 14ab + 10ab^2$

16. $25x^4 + 10x - 20x^2$

17. $bx^3 - b^2x^2 + b^3x$

18. $4k^3a^2 + 22ka + 16k^2a^2$

19. $33w^4y^2 - 9w^3y^2 + 24w^2y^2$

20. $18x^3 - 9x^5 + 27x^2$

## 6.2   Finding the Numbers

The next kind of factoring we will do requires thinking of two numbers with a certain sum and a certain product.

**Example 4:**   Which two numbers have a sum of 8 and a product of 12? In other words, what pair of numbers would answer both equations?

$$\underline{\hspace{1cm}} + \underline{\hspace{1cm}} = 8 \quad \text{and} \quad \underline{\hspace{1cm}} \times \underline{\hspace{1cm}} = 12$$

You may think $4 + 4 = 8$, but $4 \times 4$ does not equal 12.
Or you may think $7 + 1 = 8$, but $7 \times 1$ does not equal 12.

$6 + 2 = 8$ and $6 \times 2 = 12$, so 6 and 2 are the pair of numbers that will work in both equations.

**For each problem below, find one pair of numbers that will solve both equations.**

1.  $\underline{\hspace{1cm}} + \underline{\hspace{1cm}} = 14$   and   $\underline{\hspace{1cm}} \times \underline{\hspace{1cm}} = 40$

2.  $\underline{\hspace{1cm}} + \underline{\hspace{1cm}} = 10$   and   $\underline{\hspace{1cm}} \times \underline{\hspace{1cm}} = 21$

3.  $\underline{\hspace{1cm}} + \underline{\hspace{1cm}} = 18$   and   $\underline{\hspace{1cm}} \times \underline{\hspace{1cm}} = 81$

4.  $\underline{\hspace{1cm}} + \underline{\hspace{1cm}} = 12$   and   $\underline{\hspace{1cm}} \times \underline{\hspace{1cm}} = 20$

5.  $\underline{\hspace{1cm}} + \underline{\hspace{1cm}} = 7$   and   $\underline{\hspace{1cm}} \times \underline{\hspace{1cm}} = 12$

6.  $\underline{\hspace{1cm}} + \underline{\hspace{1cm}} = 8$   and   $\underline{\hspace{1cm}} \times \underline{\hspace{1cm}} = 15$

7.  $\underline{\hspace{1cm}} + \underline{\hspace{1cm}} = 10$   and   $\underline{\hspace{1cm}} \times \underline{\hspace{1cm}} = 25$

8.  $\underline{\hspace{1cm}} + \underline{\hspace{1cm}} = 14$   and   $\underline{\hspace{1cm}} \times \underline{\hspace{1cm}} = 48$

9.  $\underline{\hspace{1cm}} + \underline{\hspace{1cm}} = 12$   and   $\underline{\hspace{1cm}} \times \underline{\hspace{1cm}} = 36$

10.  $\underline{\hspace{1cm}} + \underline{\hspace{1cm}} = 17$   and   $\underline{\hspace{1cm}} \times \underline{\hspace{1cm}} = 72$

11.  $\underline{\hspace{1cm}} + \underline{\hspace{1cm}} = 15$   and   $\underline{\hspace{1cm}} \times \underline{\hspace{1cm}} = 56$

12.  $\underline{\hspace{1cm}} + \underline{\hspace{1cm}} = 9$   and   $\underline{\hspace{1cm}} \times \underline{\hspace{1cm}} = 18$

13.  $\underline{\hspace{1cm}} + \underline{\hspace{1cm}} = 13$   and   $\underline{\hspace{1cm}} \times \underline{\hspace{1cm}} = 40$

14.  $\underline{\hspace{1cm}} + \underline{\hspace{1cm}} = 16$   and   $\underline{\hspace{1cm}} \times \underline{\hspace{1cm}} = 63$

15.  $\underline{\hspace{1cm}} + \underline{\hspace{1cm}} = 10$   and   $\underline{\hspace{1cm}} \times \underline{\hspace{1cm}} = 16$

16.  $\underline{\hspace{1cm}} + \underline{\hspace{1cm}} = 8$   and   $\underline{\hspace{1cm}} \times \underline{\hspace{1cm}} = 16$

17.  $\underline{\hspace{1cm}} + \underline{\hspace{1cm}} = 9$   and   $\underline{\hspace{1cm}} \times \underline{\hspace{1cm}} = 20$

18.  $\underline{\hspace{1cm}} + \underline{\hspace{1cm}} = 13$   and   $\underline{\hspace{1cm}} \times \underline{\hspace{1cm}} = 36$

19.  $\underline{\hspace{1cm}} + \underline{\hspace{1cm}} = 15$   and   $\underline{\hspace{1cm}} \times \underline{\hspace{1cm}} = 50$

20.  $\underline{\hspace{1cm}} + \underline{\hspace{1cm}} = 11$   and   $\underline{\hspace{1cm}} \times \underline{\hspace{1cm}} = 30$

# 6.3   More Finding the Numbers

Now that you have mastered positive numbers, take up the challenge of finding pairs of negative numbers or pairs where one number is negative and one is positive.

**Example 5:**   Which two numbers have a sum of $-3$ and a product of $-40$? In other words, what pair of numbers would answer both equations?

$$\underline{\hspace{3em}} + \underline{\hspace{3em}} = -3 \quad \text{and} \quad \underline{\hspace{3em}} \times \underline{\hspace{3em}} = -40$$

It is faster to look at the factors of 40 first. 8 and 5 and 10 and 4 are possibilities. 8 and 5 have a difference of 3, and in fact, $5 + (-8) = -3$ and $5 \times (-8) = -40$. This pair of numbers, 5 and $-8$, will satisfy both equations.

**For each problem below, find one pair of numbers that will solve both equations.**

1.   $\underline{\hspace{3em}} + \underline{\hspace{3em}} = -2$   and   $\underline{\hspace{3em}} \times \underline{\hspace{3em}} = -35$

2.   $\underline{\hspace{3em}} + \underline{\hspace{3em}} = 4$   and   $\underline{\hspace{3em}} \times \underline{\hspace{3em}} = -5$

3.   $\underline{\hspace{3em}} + \underline{\hspace{3em}} = 4$   and   $\underline{\hspace{3em}} \times \underline{\hspace{3em}} = -12$

4.   $\underline{\hspace{3em}} + \underline{\hspace{3em}} = -6$   and   $\underline{\hspace{3em}} \times \underline{\hspace{3em}} = 8$

5.   $\underline{\hspace{3em}} + \underline{\hspace{3em}} = 3$   and   $\underline{\hspace{3em}} \times \underline{\hspace{3em}} = -40$

6.   $\underline{\hspace{3em}} + \underline{\hspace{3em}} = 10$   and   $\underline{\hspace{3em}} \times \underline{\hspace{3em}} = -11$

7.   $\underline{\hspace{3em}} + \underline{\hspace{3em}} = 6$   and   $\underline{\hspace{3em}} \times \underline{\hspace{3em}} = -27$

8.   $\underline{\hspace{3em}} + \underline{\hspace{3em}} = 8$   and   $\underline{\hspace{3em}} \times \underline{\hspace{3em}} = -20$

9.   $\underline{\hspace{3em}} + \underline{\hspace{3em}} = -5$   and   $\underline{\hspace{3em}} \times \underline{\hspace{3em}} = -24$

10.   $\underline{\hspace{3em}} + \underline{\hspace{3em}} = -3$   and   $\underline{\hspace{3em}} \times \underline{\hspace{3em}} = -28$

11.   $\underline{\hspace{3em}} + \underline{\hspace{3em}} = -2$   and   $\underline{\hspace{3em}} \times \underline{\hspace{3em}} = -48$

12.   $\underline{\hspace{3em}} + \underline{\hspace{3em}} = -1$   and   $\underline{\hspace{3em}} \times \underline{\hspace{3em}} = -20$

13.   $\underline{\hspace{3em}} + \underline{\hspace{3em}} = -3$   and   $\underline{\hspace{3em}} \times \underline{\hspace{3em}} = 2$

14.   $\underline{\hspace{3em}} + \underline{\hspace{3em}} = 1$   and   $\underline{\hspace{3em}} \times \underline{\hspace{3em}} = -30$

15.   $\underline{\hspace{3em}} + \underline{\hspace{3em}} = -7$   and   $\underline{\hspace{3em}} \times \underline{\hspace{3em}} = 12$

16.   $\underline{\hspace{3em}} + \underline{\hspace{3em}} = 6$   and   $\underline{\hspace{3em}} \times \underline{\hspace{3em}} = -16$

17.   $\underline{\hspace{3em}} + \underline{\hspace{3em}} = 5$   and   $\underline{\hspace{3em}} \times \underline{\hspace{3em}} = -24$

18.   $\underline{\hspace{3em}} + \underline{\hspace{3em}} = -4$   and   $\underline{\hspace{3em}} \times \underline{\hspace{3em}} = 4$

19.   $\underline{\hspace{3em}} + \underline{\hspace{3em}} = -1$   and   $\underline{\hspace{3em}} \times \underline{\hspace{3em}} = -42$

20.   $\underline{\hspace{3em}} + \underline{\hspace{3em}} = -6$   and   $\underline{\hspace{3em}} \times \underline{\hspace{3em}} = 8$

## 6.4   Factoring Trinomials

In the chapter on polynomials, you multiplied binomials (two terms) together, and the answer was a trinomial (three terms).

For example, $(x+6)(x-5) = x^2 + x - 30$

Now, you need to practice factoring a trinomial into two binomials.

**Example 6:**   Factor $x^2 + 6x + 8$

**Step 1:**   When the trinomial is in descending order as in the example above, you need to find a pair of numbers whose sum equals the number in the second term, while their product equals the third term. In the above example, find the pair of numbers that has a sum of 6 and a product of 8.

$$\underline{\hspace{1cm}} + \underline{\hspace{1cm}} = 6 \quad \text{and} \quad \underline{\hspace{1cm}} \times \underline{\hspace{1cm}} = 8$$

The pair of numbers that satisfy both equations is 4 and 2.

**Step 2:**   Use the pair of numbers in the binomials.

The factors of $x^2 + 6x + 8$ are $(x+4)(x+2)$

**Check:**   To check, use the FOIL method.
$(x+4)(x+2) = x^2 + 4x + 2x + 8 = x^2 + 6x + 8$

**Notice, when the second term and the third term of the trinomial are both positive, both numbers in the solution are positive.**

**Example 7:**   Factor $x^2 - x - 6$        Find the pair of numbers where:

the sum is $-1$ and the product is $-6$

$$\underline{\hspace{1cm}} + \underline{\hspace{1cm}} = -1 \quad \text{and} \quad \underline{\hspace{1cm}} \times \underline{\hspace{1cm}} = -6$$

The pair of numbers that satisfies both equations is 2 and $-3$.
The factors of $x^2 - x - 6$ are $(x+2)(x-3)$

**Notice, if the second term and the third term are negative, one number in the solution pair is positive, and the other number is negative.**

**Example 8:**    Factor $x^2 - 7x + 12$         Find the pair of numbers where:

the sum is $-7$ and the product is 12

_____ + _____ = $-7$    and    _____ × _____ = 12

The pair of numbers that satisfies both equations is $-3$ and $-4$
The factors of $x^2 - 7x + 12$ are $(x - 3)(x - 4)$.

**Notice, if the second term of a trinomial is negative and the third term is positive, both numbers in the solution are negative.**

**Find the factors of the following trinomials.**

1. $x^2 - x - 2$

2. $y^2 + y - 6$

3. $w^2 + 3w - 4$

4. $t^2 + 5t + 6$

5. $x^2 + 2x - 8$

6. $k^2 - 4k + 3$

7. $t^2 + 3t - 10$

8. $x^2 - 3x - 4$

9. $y^2 - 5y + 6$

10. $y^2 + y - 20$

11. $a^2 - a - 6$

12. $b^2 - 4b - 5$

13. $c^2 - 5c - 14$

14. $c^2 - c - 12$

15. $d^2 + d - 6$

16. $x^2 - 3x - 28$

17. $y^2 + 3y - 18$

18. $a^2 - 9a + 20$

19. $b^2 - 2b - 15$

20. $c^2 + 7c - 8$

21. $t^2 - 11t + 30$

22. $w^2 + 13w + 36$

23. $m^2 - 2m - 48$

24. $y^2 + 14y + 49$

25. $x^2 + 7x + 10$

26. $a^2 - 7a + 6$

27. $d^2 - 6d - 27$

## 6.5   More Factoring Trinomials

Sometimes a trinomial has a greatest common factor which must be factored out first.

**Example 9:**       Factor $4x^2 + 8x - 32$

**Step 1:**       Begin by factoring out the greatest common factor, 4.

$4\left(x^2 + 2x - 8\right)$

**Step 2:**       Factor by finding a pair of numbers whose sum is 2 and product is $-8$. 4 and $-2$ will work, so

$4\left(x^2 + 2x - 8\right) = 4\left(x + 4\right)\left(x - 2\right)$

**Check:**       Multiply to check. $4\left(x + 4\right)\left(x - 2\right) = 4x^2 + 8x - 32$

**Factor the following trinomials. Be sure to factor out the greatest common factor first.**

1. $2x^2 + 6x + 4$

2. $3y^2 - 9y + 6$

3. $2a^2 + 2a - 12$

4. $4b^2 + 28b + 40$

5. $3y^2 - 6y - 9$

6. $10x^2 + 10x - 200$

7. $5c^2 - 10c - 40$

8. $6d^2 + 30d - 36$

9. $4x^2 + 8x - 60$

10. $6a^2 - 18a - 24$

11. $5b^2 + 40b + 75$

12. $3c^2 - 6c - 24$

13. $2x^2 - 18x + 28$

14. $4y^2 - 20y + 16$

15. $7a^2 - 7a - 42$

16. $6b^2 - 18b - 60$

17. $11d^2 + 66d + 88$

18. $3x^2 - 24x + 45$

# 6.6   Factoring More Trinomials

Some trinomials have a whole number in front of the first term that cannot be factored out of the trinomial. The trinomial can still be factored.

**Example 10:**    Factor $2x^2 + 5x - 3$

**Step 1:**    To get a product of $2x^2$, one factor must begin with $2x$ and the other with $x$.

$$(2x \quad)(x \quad)$$

**Step 2:**    Now think: What two numbers give a product of $-3$? The two possibilities are 3 and $-1$ or $-3$ and 1. We know they could be in any order so there are 4 possible arrangements.

$$(2x + 3)(x - 1)$$
$$(2x - 3)(x + 1)$$
$$(2x + 1)(x - 3)$$
$$(2x - 1)(x + 3)$$

**Step 3:**    Multiply each possible answer until you find the arrangement of the numbers that works. Multiply the outside terms and the inside terms and add them together to see which one will equal $5x$.

$$(2x + 3)(x - 1) = 2x^2 + x - 3$$
$$(2x - 3)(x + 1) = 2x^2 - x - 3$$
$$(2x + 1)(x - 3) = 2x^2 - 5 - 3$$
$$\boxed{(2x - 1)(x + 3) = 2x^2 + 5x - 3}$$ $\longleftarrow$ This arrangement works, therefore:

The factors of $2x^2 + 5x - 3$ are $(2x - 1)(x + 3)$

**Alternative:**    You can do some of the multiplying in your head. For the above example, ask yourself the following question: What two numbers give a product of $-3$ and give a sum of 5 (the whole number in the second term) when one number is first multiplied by 2 (the whole number in front of the first term)? The pair of numbers, $-1$ and 3, have a product of $-3$ and a sum of 5 when the 3 is first multiplied by 2. Therefore, the 3 will go in the opposite factor of the $2x$ so that when the terms are multiplied, you get $-5$.

You can use this method to at least narrow down the possible pairs of numbers when you have several from which to choose.

**Factor the following trinomials.**

1. $3y^2 + 14y + 8$

2. $5a^2 + 24a - 5$

3. $7b^2 + 30b + 8$

4. $2c^2 - 9c + 9$

5. $2y^2 - 7y - 15$

6. $3x^2 + 4x + 1$

7. $7y^2 + 13y - 2$

8. $11a^2 + 35a + 6$

9. $5y^2 + 17y - 12$

10. $3a^2 + 4a - 7$

11. $2a^2 + 3a - 20$

12. $5b^2 - 13b - 6$

13. $3y^2 - 4y - 32$

14. $2x^2 - 17x + 36$

15. $11x^2 - 29x - 12$

16. $5c^2 + 2c - 16$

17. $7y^2 - 30y + 27$

18. $2x^2 - 3x - 20$

19. $5b^2 + 24b - 5$

20. $7d^2 + 18d + 8$

21. $3x^2 - 20x + 25$

22. $2a^2 - 7a - 4$

23. $5m^2 + 12m + 4$

24. $9y^2 - 5y - 4$

25. $2b^2 - 13b + 18$

26. $7x^2 + 31x - 20$

27. $3c^2 - 2c - 21$

## 6.7   Factoring the Difference of Two Squares

Let's give an example of a **perfect square**.

25 is a perfect square because $5 \times 5 = 25$

49 is a perfect square because $7 \times 7 = 49$

Any variable with an even exponent is a perfect square.

$y^2$ is a perfect square because $y \times y = y^2$

$y^4$ is a perfect square because $y^2 \times y^2 = y^4$

When two terms that are both perfect squares are subtracted, factoring those terms is very easy. To factor the difference of perfect squares, you use the square root of each term, a plus sign in the first factor, and a minus sign in the second factor.

**Example 11:**   Factor $4x^2 - 9$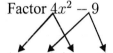

This example has two terms which are both perfect squares, and the terms are subtracted.

**Step 1:**   $(2x \quad 3)(2x \quad 3)$

Find the square root of each term.
Use the square roots in each of the factors.

**Step 2:**   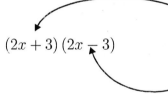 $(2x + 3)(2x - 3)$

Use a plus sign in one factor and a minus sign in the other factor.

**Check:**   Multiply to check. $(2x + 3)(2x - 3) = 4x^2 - 6x + 6x - 9 = 4x^2 - 9$

The inner and outer terms add to zero.

**Example 12:**   Factor $81y^4 - 1$

**Step 1:**   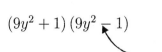 $(9y^2 + 1)(9y^2 - 1)$

Factor like the example above.
Notice, the second factor is also the difference of two perfect squares.

**Step 2:**   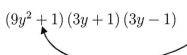 $(9y^2 + 1)(3y + 1)(3y - 1)$

Factor the second term further.
**Note: You cannot factor the sum of two perfect squares.**

**Check:**   Multiply in reverse to check your answer.

$(9y^2 + 1)(3y + 1)(3y - 1) = (9y^2 + 1)(9y^2 - 3y + 3y - 1) =$

$(9y^2 + 1)(9y^2 - 1) = 81y^4 + 9y^2 - 9y^2 - 1 = 81y^4 - 1$

**Factor the following differences of perfect squares.**

1. $64x^2 - 49$

2. $4y^4 - 25$

3. $9a^4 - 4$

4. $25c^4 - 9$

5. $64y^2 - 9$

6. $x^4 - 16$

7. $49x^2 - 4$

8. $4d^2 - 25$

9. $9a^2 - 16$

10. $100y^4 - 49$

11. $c^4 - 36$

12. $36x^2 - 25$

13. $25x^2 - 4$

14. $9x^4 - 64$

15. $49x^2 - 100$

16. $16x^2 - 81$

17. $9y^4 - 1$

18. $49c^2 - 25$

19. $25d^2 - 64$

20. $36a^4 - 49$

21. $16x^4 - 16$

22. $b^2 - 25$

23. $c^4 - 144$

24. $9y^2 - 4$

25. $81x^4 - 16$

26. $4b^2 - 36$

27. $9w^2 - 9$

28. $64a^2 - 25$

29. $49y^2 - 121$

30. $x^6 - 9$

## 6.8   Simplifying Algebraic Ratios

We will use what we learned so far in this chapter to factor the terms in the numerator and the denominator when possible, then simplify the algebraic ratio.

**Example 13:**   Simplify $\dfrac{c^2 - 25}{c^2 + 5c}$

**Step 1:**   The numerator is the difference of two perfect squares, so it can be easily factored as in the previous section. Use the square root of each of the terms in the parentheses, with a plus sign in one and a minus sign in the other.
$$c^2 - 25 = (c - 5)(c + 5)$$

**Step 2:**   Find the greatest common factor in the denominator and factor it out. In this case, it is the variable $c$.
$$c^2 + 5c = c(c + 5)$$

**Step 3:**   Simplify $\dfrac{c^2 - 25}{c^2 + 5c} = \dfrac{(c - 5)(c + 5)}{c(c + 5)} = \dfrac{c - 5}{c}$

**Simplify the algebraic ratios. Check for perfect squares and common factors.**

1. $\dfrac{25x^2 - 4}{5x^2 - 2x}$

2. $\dfrac{64c^2 - 25}{8c^2 + 5c}$

3. $\dfrac{36a^2 - 49}{6a^2 - 7a}$

4. $\dfrac{x^2 - 9}{x^2 + 3x}$

5. $\dfrac{9a^2 - 16}{3a^2 - 4a}$

6. $\dfrac{16x^2 - 81}{4x^2 - 9x}$

7. $\dfrac{49x^2 - 100}{7x^2 + 10x}$

8. $\dfrac{x^4 - 16}{x^2 + 2x}$

9. $\dfrac{4y^2 - 36}{2y^2 + 6y}$

10. $\dfrac{81y^4 - 16}{9y^2 + 4}$

11. $\dfrac{25x^4 - 225}{5x^2 + 15}$

12. $\dfrac{3y^3 + 9}{y^9 - 9}$

## 6.9    Solving Algebraic Proportions

Problems of the form,

$$\frac{ax + b}{c} = \frac{dx + e}{f}$$

may be called **algebraic proportions**. The objective is to solve for $x$ using the rules of algebra. Multiplying both sides by the denominators can be a convenient time-saver. Even better, you might notice that instead of multiplying each denominator one at a time, you may multiply both sides of the equation by the least common multiple of the denominators to help quickly solve for $x$.

**Example 14:**    Solve the algebraic proportion. $\dfrac{x + 10}{3} = \dfrac{3x + 2}{5}$

**Step 1:**    Multiply each side by 3. $x + 10 = \dfrac{3(3x + 2)}{5}$

**Step 2:**    Multiply each side by 5, which gives $5(x + 10) = 3(3x + 2)$. Notice that we could have just multiplied through by 15, the least common multiple of 3 and 5, to arrive at this stage.

**Step 3:**    Distribute. $5x + 50 = 9x + 6$

**Step 4:**    Solve for $x$.

$$
\begin{array}{rcl}
5x + 50 & = & 9x + 6 \\
-5x & & -5x \quad \text{Subtract } 5x \text{ from each side.} \\
\hline
50 & = & 4x + 6 \\
-6 & & -6 \quad \text{Subtract 6 from each side.} \\
\hline
44 & = & 4x \quad \text{Divide each side by 4.} \\
11 & = & x
\end{array}
$$

**Solve for $x$.**

1. $\dfrac{-6x - 6}{9} = \dfrac{-4x - 4}{4}$

2. $\dfrac{-8x - 8}{5} = \dfrac{-5x - 5}{4}$

3. $\dfrac{5x - 5}{6} = \dfrac{4 - 7x}{6}$

4. $\dfrac{3 - 7x}{6} = \dfrac{9x + 1}{9}$

5. $\dfrac{7x - 1}{2} = \dfrac{10x - 3}{6}$

6. $\dfrac{x - 7}{7} = -\dfrac{3}{8}$

7. $\dfrac{10x + 2}{9} = \dfrac{10 - 10x}{10}$

8. $\dfrac{3x - 10}{5} = \dfrac{-6x - 3}{10}$

9. $\dfrac{x + 4}{4} = \dfrac{7x + 8}{7}$

10. $\dfrac{3x + 5}{9} = \dfrac{8x + 9}{4}$

11. $\dfrac{x - 2}{3} = \dfrac{-x - 2}{2}$

12. $\dfrac{-5x - 2}{4} = \dfrac{8 - 7x}{5}$

13. $\dfrac{3x + 3}{4} = \dfrac{2x + 2}{5}$

14. $\dfrac{6x + 5}{9} = \dfrac{4x + 4}{8}$

15. $\dfrac{3x + 8}{3} = 9 - x$

# Chapter 6 Review

**Factor the following polynomials completely.**

1. $8x - 18$

10. $12b^2 + 25b - 7$

2. $6x^2 - 18x$

11. $c^2 + cd - 20d^2$

3. $16b^3 + 8b$

12. $6y^2 + 30y + 36$

4. $15a^3 + 40$

13. $2b^2 + 6b - 20$

5. $20y^6 - 12y^4$

14. $16b^4 - 81d^4$

6. $5a - 15a^2$

15. $9w^2 - 54w - 63$

7. $4y^2 - 36$

16. $12x^2 + 27x$

8. $2b^2 - 2b - 12$

17. $2a^4 - 32$

9. $27y^2 + 42y - 5$

18. $21c^2 + 41c + 10$

**Solve for $x$.**

19. $\dfrac{-x + 11}{8} = \dfrac{3x - 2}{5}$

21. $\dfrac{-8x}{5} = -4x - 3$

20. $\dfrac{4 - x}{4} = \dfrac{9x + 4}{2}$

22. $\dfrac{x - 2}{4} = 6 - 6x$

## Chapter 6 Test

1. What is the greatest common factor of $4x^3$ and $8x^2$?

   **A** $4x^2$
   **B** $4x$
   **C** $x^2$
   **D** $8x$

2. Factor: $8x^4 - 7x^2 + 4x$

   **A** $4x\left(2x^3 - 7x + 4\right)$
   **B** $x\left(8x^4 - 7x^2 + 4x\right)$
   **C** $x\left(8x^3 - 7x + 4\right)$
   **D** $4x\left(2x^3 - 7x + 1\right)$

3. Factor: $x^2 + 6x + 8$

   **A** $(x + 2)(x + 4)$
   **B** $(x + 1)(x + 8)$
   **C** $(x - 2)(x - 4)$
   **D** $(x - 1)(x - 8)$

4. Simplify the following algebraic ratio:
   $$\frac{36x^4 - 16}{6x^3 + 4x}$$

   **A** $\dfrac{\left(6x^2 - 4\right)\left(6x^2 + 4\right)}{x}$

   **B** $\dfrac{6x^2 - 4}{x}$

   **C** $6x^2 - 4$

   **D** $\dfrac{(3x - 2)(3x + 2)}{x}$

5. Factor: $2x^2 - 2x - 84$

   **A** $(2x + 7)(x - 12)$
   **B** $(2x - 12)(x + 7)$
   **C** $(2x - 7)(x + 12)$
   **D** $(2x + 12)(x - 7)$

6. Simplify the following algebraic ratio:
   $$\frac{c^2 + 10c + 24}{c^3 + 4c^2}$$

   **A** $\dfrac{c + 6}{c}$

   **B** $\dfrac{c + 4}{c^2}$

   **C** $\dfrac{c + 6}{c^2}$

   **D** $\dfrac{(c + 6)(c + 4)}{c^2(c + 4)}$

7. Factor: $4x^2 - 64$

   **A** $(x - 8)(x + 8)$
   **B** $(4x - 8)(4x + 8)$
   **C** $(2x - 16)(2x + 16)$
   **D** $(2x - 8)(2x + 8)$

8. Solve: $6x - 5 = \dfrac{6x - 4}{2}$

   **A** $x = 1$

   **B** $x = -\dfrac{7}{9}$

   **C** $x = -1$

   **D** $x = \dfrac{5}{18}$

9. Solve: $\dfrac{2x + 7}{2} = \dfrac{8x - 1}{6}$

   **A** $x = -11$

   **B** $x = -\dfrac{10}{7}$

   **C** $x = 11$

   **D** $x = \dfrac{10}{7}$

# Chapter 7
# Solving Quadratic Equations

This chapter covers the following IN Algebra I standards:

| Standard 6: | Polynomials | A1.6.8 |
|---|---|---|
| Standard 8: | Quadratic, Cubic, and Radical | A1.8.2 |
| | Equations | A1.8.3 |
| | | A1.8.4 |
| | | A1.8.5 |
| | | A1.8.6 |
| | | A1.8.7 |
| | | A1.8.8 |
| | | A1.8.9 |

In the previous chapter, we factored polynomials such as $y^2 - 4y - 5$ into two factors:

$$y^2 - 4y - 5 = (y+1)(y-5)$$

In this chapter, we learn that any equation that can be put in the form $ax^2 + bx + c = 0$ is a quadratic equation if $a$, $b$, and $c$ are real numbers and $a \neq 0$. $ax^2 + bx + c = 0$ is the standard form of a quadratic equation. To solve these equations, follow the steps below.

**Example 1:** Solve $y^2 - 4y - 5 = 0$

**Step 1:** Factor the left side of the equation.

$$
\begin{aligned}
y^2 - 4y - 5 &= 0 \\
(y+1)(y-5) &= 0
\end{aligned}
$$

**Step 2:** If the product of these two factors equals zero, then the two factors individually must be equal to zero. Therefore, to solve, we set each factor equal to zero.

$$
\begin{array}{ll}
(y+1) = 0 & \quad (y-5) = 0 \\
\underline{-1 \quad -1} & \quad \underline{+5 \quad +5} \\
y = -1 & \quad y = 5
\end{array}
$$

The equation has two solutions: $y = -1$ and $y = 5$

**Check:** To check, substitute each solution into the original equation.

When $y = -1$, the equation becomes:     When $y = 5$, the equation becomes:

$$
\begin{array}{ll}
(-1)^2 - (4)(-1) - 5 = 0 & \quad 5^2 - (4)(5) - 5 = 0 \\
1 + 4 - 5 = 0 & \quad 25 - 20 - 5 = 0 \\
0 = 0 & \quad 0 = 0
\end{array}
$$

Both solutions produce true statements.
The solution set for the equation is $\{-1, -5\}$.

**Solve each of the following quadratic equations by factoring and setting each factor equal to zero. Check by substituting answers back in the original equation.**

1. $x^2 + x - 6 = 0$

2. $y^2 - 2y - 8 = 0$

3. $a^2 + 2a - 15 = 0$

4. $y^2 - 5y + 4 = 0$

5. $b^2 - 9b + 14 = 0$

6. $x^2 - 3x - 4 = 0$

7. $y^2 + y - 20 = 0$

8. $d^2 + 6d + 8 = 0$

9. $y^2 - 7y + 12 = 0$

10. $x^2 - 3x - 28 = 0$

11. $a^2 - 5a + 6 = 0$

12. $b^2 + 3b - 10 = 0$

13. $a^2 + 7a - 8 = 0$

14. $x^2 + 3x + 2 = 0$

15. $x^2 - x - 42 = 0$

16. $a^2 + a - 6 = 0$

17. $b^2 + 7b + 12 = 0$

18. $y^2 + 2y - 15 = 0$

19. $a^2 - 3a - 10 = 0$

20. $d^2 + 10d + 16 = 0$

21. $x^2 - 4x - 12 = 0$

Quadratic equations that have a whole number and a variable in the first term are solved the same way as the previous page. Factor the trinomial, and set each factor equal to zero to find the solution set.

**Example 2:**   Solve $2x^2 + 3x - 2 = 0$

$(2x - 1)(x + 2) = 0$

Set each factor equal to zero and solve:

$$2x - 1 = 0$$
$$\underline{+1 \quad +1}$$
$$\frac{2x}{2} = \frac{1}{2}$$
$$x = \frac{1}{2}$$

$$x + 2 = 0$$
$$\underline{-2 \quad -2}$$
$$x = -2$$

The solution set is $\left\{ \dfrac{1}{2}, -2 \right\}$.

**Solve the following quadratic equations.**

22. $3y^2 + 4y - 32 = 0$

23. $5c^2 - 2c - 16 = 0$

24. $7d^2 + 18d + 8 = 0$

25. $3a^2 - 10a - 8 = 0$

26. $11x^2 - 31x - 6 = 0$

27. $5b^2 + 17b + 6 = 0$

28. $3x^2 - 11x - 20 = 0$

29. $5a^2 + 47a - 30 = 0$

30. $2c^2 - 5c - 25 = 0$

31. $2y^2 + 11y - 21 = 0$

32. $5a^2 + 23a - 42 = 0$

33. $3d^2 + 11d - 20 = 0$

34. $3x^2 - 10x + 8 = 0$

35. $7b^2 + 23b - 20 = 0$

36. $9a^2 - 58a + 24 = 0$

37. $4c^2 - 25c - 21 = 0$

38. $8d^2 + 53d + 30 = 0$

39. $4y^2 + 37y - 30 = 0$

40. $8a^2 + 37a - 15 = 0$

41. $3x^2 - 41x + 26 = 0$

42. $8b^2 + 2b - 3 = 0$

## 7.1 Solving Perfect Squares

When the square root of a constant, variable, or polynomial results in a constant, variable, or polynomial without irrational numbers, the expression is a **perfect square**. Some examples are 49, $x^2$, and $(x-2)^2$.

**Example 3:** Solve the perfect square for $x$. $(x-5)^2 = 0$

**Step 1:** Take the square root of both sides.
$$\sqrt{(x-5)^2} = \sqrt{0}$$
$$(x-5) = 0$$

**Step 2:** Solve the equation.
$$(x-5) = 0$$
$$x - 5 + 5 = 0 + 5$$
$$x = 5$$

**Example 4:** Solve the perfect square for $x$. $(x-5)^2 = 64$

**Step 1:** Take the square root of both sides.
$$\sqrt{(x-5)^2} = \sqrt{64}$$
$$(x-5) = \pm 8$$
$$(x-5) = 8 \text{ and } (x-5) = -8$$

**Step 2:** Solve the two equations.
$$(x-5) = 8 \qquad \text{and} \quad (x-5) = -8$$
$$x - 5 + 5 = 8 + 5 \quad \text{and} \quad x - 5 + 5 = -8 + 5$$
$$x = 13 \qquad\qquad \text{and} \quad x = -3$$

**Solve the perfect square for $x$.**

1. $(x-2)^2 = 0$

2. $(x+1)^2 = 0$

3. $(x+11)^2 = 0$

4. $(x-4)^2 = 0$

5. $(x-1)^2 = 0$

6. $(x+8)^2 = 0$

7. $(x+3)^2 = 4$

8. $(x-5)^2 = 16$

9. $(x-10)^2 = 100$

10. $(x+9)^2 = 9$

11. $(x-4.5)^2 = 25$

12. $(x+7)^2 = 36$

13. $(x+2)^2 = 49$

14. $(x-1)^2 = 4$

15. $(x+8.9)^2 = 49$

16. $(x-6)^2 = 81$

17. $(x-12)^2 = 121$

18. $(x+2.5)^2 = 64$

## 7.2   Completing the Square

"Completing the Square" is another way of factoring a quadratic equation. To complete the square, convert the equation into a perfect square.

**Example 5:**   Solve $x^2 - 10x + 9 = 0$ by completing the square.

**Completing the square:**

**Step 1:**   The first step is to get the constant on the other side of the equation. Subtract 9 from both sides:
$$x^2 - 10x + 9 - 9 = -9$$
$$x^2 - 10x = -9$$

**Step 2:**   Determine the coefficient of the $x$. The coefficient in this example is 10. Divide the coefficient by 2 and square the result.
$$(10 \div 2)^2 = 5^2 = 25$$

**Step 3:**   Add the resulting value, 25, to both sides:
$$x^2 - 10x + 25 = -9 + 25$$
$$x^2 - 10x + 25 = 16$$

**Step 4:**   Now factor the $x^2 - 10x + 25$ into a perfect square:
$$(x - 5)^2 = 16$$

**Solving the perfect square:**

**Step 5:**   Take the square root of both sides.
$$\sqrt{(x - 5)^2} = \sqrt{16}$$
$$(x - 5) = \pm 4$$
$$(x - 5) = 4 \text{ and } (x - 5) = -4$$

**Step 6:**   Solve the two equations.
$$\begin{array}{lll} (x - 5) = 4 & \text{and} & (x - 5) = -4 \\ x - 5 + 5 = 4 + 5 & \text{and} & x - 5 + 5 = -4 + 5 \\ x = 9 & \text{and} & x = 1 \end{array}$$

**Solve for $x$ by completing the square.**

1. $x^2 + 2x - 3 = 0$

2. $x^2 - 8x + 7 = 0$

3. $x^2 + 6x - 7 = 0$

4. $x^2 - 16x - 36 = 0$

5. $x^2 - 14x + 49 = 0$

6. $x^2 - 4x = 0$

7. $x^2 + 12x + 27 = 0$

8. $x^2 + 2x - 24 = 0$

9. $x^2 + 12x - 85 = 0$

10. $x^2 - 8x + 15 = 0$

11. $x^2 - 16x + 60 = 0$

12. $x^2 - 8x - 48 = 0$

13. $x^2 + 24x + 44 = 0$

14. $x^2 + 6x + 5 = 0$

15. $x^2 - 11x + 5.25 = 0$

## 7.3 Proof of the Quadratic Formula

The quadratic formula $\dfrac{-b \pm \sqrt{b^2 - 4ac}}{2a}$ can be proved by using the "completing the square" method on the quadratic equation $ax^2 + bx + c = 0$.

$$ax^2 + bx + c = 0$$

$$ax^2 + bx + c - c = 0 - c \qquad \text{Subtract } c \text{ from both sides.}$$

$$ax^2 + bx = -c \qquad \text{Simplify.}$$

$$\frac{ax^2 + bx}{a} = \frac{-c}{a} \qquad \text{Divide by } a \text{ on both sides.}$$

$$x^2 + \frac{bx}{a} = \frac{-c}{a} \qquad \text{Simplify.}$$

$$x^2 + \frac{bx}{a} + \left(\frac{b}{2a}\right)^2 = \frac{-c}{a} + \left(\frac{b}{2a}\right)^2 \qquad \text{Complete the square by adding } \left(\frac{b}{2a}\right)^2 \text{ to both sides.}$$

$$\left(x + \frac{b}{2a}\right)^2 = \frac{-c}{a} + \left(\frac{b}{2a}\right)^2 \qquad \text{Factor the left side.}$$

$$\left(x + \frac{b}{2a}\right)^2 = \frac{-c}{a} + \frac{b^2}{4a^2} \qquad \text{Square } \frac{b}{2a} \text{ on the right side.}$$

$$\left(x + \frac{b}{2a}\right)^2 = \frac{-4ac}{4a^2} + \frac{b^2}{4a^2} \qquad \text{Find a common denominator on the right side so the fractions can be added.}$$

$$\left(x + \frac{b}{2a}\right)^2 = \frac{b^2 - 4ac}{4a^2} \qquad \text{Add the fractions on the right side.}$$

$$\sqrt{\left(x + \frac{b}{2a}\right)^2} = \sqrt{\frac{b^2 - 4ac}{4a^2}} \qquad \text{Take the square root of both sides.}$$

$$x + \frac{b}{2a} = \frac{\pm\sqrt{b^2 - 4ac}}{2a} \qquad \text{Simplify.}$$

$$x + \frac{b}{2a} - \frac{b}{2a} = \frac{\pm\sqrt{b^2 - 4ac}}{2a} - \frac{b}{2a} \qquad \text{Subtract } \frac{b}{2a} \text{ from both sides.}$$

$$x = \frac{-b \pm \sqrt{b^2 - 4ac}}{2a} \qquad \text{Add the fractions. The proof is complete.}$$

## 7.4    Using the Quadratic Formula

You may be asked to use the quadratic formula to solve an algebra problem known as a **quadratic equation**. The equation should be in the form $ax^2 + bx + c = 0$.

**Example 6:**    Using the quadratic formula, find $x$ in the following equation: $x^2 - 8x = -7$.

**Step 1:**    Make sure the equation is set equal to 0.

$$x^2 - 8x + 7 = -7 + 7$$
$$x^2 - 8x + 7 = 0$$

The quadratic formula, $\dfrac{-b \pm \sqrt{b^2 - 4ac}}{2a}$, will be given to you on your formula sheet with your test.

**Step 2:**    In the formula, $a$ is the number $x^2$ is multiplied by, $b$ is the number $x$ is multiplied by and $c$ is the last term of the equation. For the equation in the example, $x^2 - 8x + 7$, $a = 1$, $b = -8$, and $c = 7$. When we look at the formula we notice a $\pm$ sign. This means that there will be two solutions to the equation, one when we use the plus sign and one when we use the minus sign. Substituting the numbers from the problem into the formula, we have:

$$\frac{8 + \sqrt{8^2 - (4)(1)(7)}}{2(1)} = 7 \qquad \text{or} \qquad \frac{8 - \sqrt{8^2 - (4)(1)(7)}}{2(1)} = 1$$

The solutions are $\{7, 1\}$

**For each of the following equations, use the quadratic formula to find two solutions.**

1. $x^2 + x - 6 = 0$

2. $y^2 - 2y - 8 = 0$

3. $a^2 + 2a - 15 = 0$

4. $y^2 - 5y + 4 = 0$

5. $b^2 - 9b + 14 = 0$

6. $x^2 - 3x - 4 = 0$

7. $y^2 + y - 20 = 0$

8. $d^2 + 6d + 8 = 0$

9. $y^2 - 7y + 12 = 0$

10. $x^2 - 3x - 28 = 0$

11. $a^2 - 5a + 6 = 0$

12. $b^2 + 3b - 10 = 0$

13. $a^2 + 7a - 8 = 0$

14. $c^2 + 3c + 2 = 0$

15. $x^2 - x - 42 = 0$

16. $a^2 + 5a - 6 = 0$

17. $b^2 + 7b + 12 = 0$

18. $y^2 + y - 12 = 0$

19. $a^2 - 3a - 10 = 0$

20. $d^2 + 10d + 16 = 0$

21. $x^2 - 4x - 12 = 0$

# 7.5   Real-World Quadratic Equations

The most common real life situation that would use a quadratic equation is the motion of an object under the force of gravity. Two examples are a ball being kicked into the air or a rocket being shot into the air.

**Example 7:**   A high school football player is practicing his field goal kicks. The equation below represents the height of the ball at a specific time.

$$s = -9t^2 + 45t$$

$t =$ amount of time in seconds

$s =$ height in feet

**Question 1:**   Where will the ball be at $4$ seconds?

**Solution 1:**   Since there are only two variables, you will only need the value of one variable to solve the problem. Simply plug in the number $4$ in place of the variable $t$ and solve the equation as shown below.

$$s = -9\,(4)^2 + 45\,(4)$$
$$s = -9\,(16) + 180$$
$$s = -144 + 180$$
$$s = 36$$

At $4$ seconds the ball will be $36$ ft in the air.

**Question 2:**   If the ball is $54$ ft in the air, how much time has gone by?

**Solution 2:**   This question is similar to the previous one, except that the given variable is different. This time you would be replacing $s$ with $54$ and then solve the equation.

| | |
|---|---|
| $54 = -9t^2 + 45t$ | Subtract $54$ on both sides. |
| $0 = -9t^2 + 45t - 54$ | Divide the entire equation by $-9$. |
| $0 = t^2 - 5t + 6$ | Factor the equation. |
| $0 = (t - 3)\,(t - 2)$ | Solve for $t$. |
| $t = 3 \qquad t = 2$ | |

For this question we got $2$ answers. The ball is $54$ ft in the air when $2$ and $3$ seconds have gone by.

**Example 8:**   John and Alex are kicking a soccer ball back and forth to each other. The equation below represents the height of the ball at a specific point in time.
$s = -4t^2 + 24t$, where $t =$ amount of time in seconds and $s =$ height in feet

**Question 1:**   How long does it take for the soccer ball to come back down to the ground?

**Solution 1:**   Looking at this problem you can see that no value was given, but one was indirectly given. The question asks when will the ball come back down. This is just another way of asking, "When will the height of the ball be zero?" The value 0 will be used for $s$. Substitute 0 back in for $s$ and solve.

$0 = -4t^2 + 24t$      Factor out the greatest common factor.
$0 = -4t(t - 6)$      Set each factor to 0 and solve.
$t = 0$      $t = 6$

It is clear that the ball is on the ground at 0 seconds, so the value of $t = 0$ is not the answer and the second value of $t$ is used instead. It takes the ball 6 seconds to go up into the air and then come back down to the ground.

**Question 2:**   What is the highest point the ball will go?

**Solution 2:**   This is asking what is the vertex of the equation. You will need to use the vertex formula. As a reminder, the quadratic equation is defined as $y = ax^2 + bx + c$, where $a \neq 0$. The quadratic equation can also be written as a function of $x$ by substituting $f(x)$ for $y$, such as $f(x) = ax^2 + bx + c$. To find the point of the vertex of the graph, you must use the formula below.

$$\text{vertex} = \left( -\frac{b}{2a}, f\left( -\frac{b}{2a} \right) \right)$$

where $f\left( -\frac{b}{2a} \right)$ is the quadratic equation evaluated at the value $-\frac{b}{2a}$. To do this, plug $-\frac{b}{2a}$ in for $x$.

To use the vertex equation, put the original equation in quadratic form and find $a$ and $b$.
$s = -4t^2 + 24t \Rightarrow -4t^2 + 24t + 0$.
Since $a$ is the coefficient of $t^2$, $a = -4$. $b$ is the coefficient of $t$, so $b = 24$.

Find the solution to $-\frac{b}{2a}$ by substituting the values of $a$ and $b$ from the equation into the expression.
$$-\frac{b}{2a} = -\left( \frac{24}{2 \times -4} \right) = -\left( \frac{24}{-8} \right) = -(-3) = 3$$

Find the solution to $f\left( -\frac{b}{2a} \right)$. We know that $-\frac{b}{2a} = 3$, so we need to find $f(3)$. To do this, we must substitute 3 into the quadratic equation for $x$.
$f(t) = -4t^2 + 24t$
$f(3) = -4(3)^2 + 24(3) = -4(9) + 72 = -36 + 72 = 36$

The vertex equals $(3, 36)$. This means at 3 seconds, the ball is 36 feet in the air. Therefore the highest the soccer ball will go is 36 ft.

**Solve the following quadratic problems.**

1. Eric is at the top of a cliff that is 500 ft from the ocean's surface. He is waiting for his friend to climb up and meet him. As he is waiting he decides to start casually tossing pebbles off the side of the cliff. The equation that represents the height of his pebbles tosses is $s = -t^2 + 5t + 500$, where $s$ = distance in feet and $t$ = time in seconds.

(A) How long does it take the pebble to hit the water?

(B) If fifteen seconds have gone by, what is the height of the pebble from the ocean?

(C) What is the highest point the pebble will go?

2. Devin is practicing golf at the driving range. The equation that represents the height of his ball is $s = -0.5t^2 + 12t$, where $s$ = distance in feet and $t$ = time in seconds.

(A) What is the highest her ball will ever go?

(B) If the ball is at $31.5$ ft in the air, how many seconds has gone by?

(C) How long will it take for the ball to hit the ground?

3. Jack throws a ball up in the air to see how high he can get it to go. The equation that represents the height of the ball is $s = -2t^2 + 20t$, where $s$ = distance in feet and $t$ = time in seconds.

(A) How high will the ball be at $7$ seconds?

(B) If the ball is $48$ feet in the air, how many seconds have gone by?

(C) How long does it take for the ball to go up and come back down to the ground?

4. Kali is jumping on her super trampoline, getting as high as she possibly can. The equation to represent her height is $s = -5t^2 + 20t$, where $s$ = distance in feet and $t$ = time in seconds.

(A) What is the highest Kali can jump on her trampoline?

(B) How high will Kali be at $4$ seconds?

(C) If Kali is $18.75$ ft in the air, then how many seconds have gone by?

## 7.6   Solving Equations That Contain Radical Expressions

Expressions such as $\sqrt{x+6}$ are considered "radical" expressions with the square root symbol being the radical. (Higher-order roots like cube roots and fourth roots are also radical expressions.) Here we will consider how to solve equations where a variable is part of the radicand (everything under the root symbol). Equations like $\sqrt{x+6} = x$.

**Example 9:**   Solve $\sqrt{x+6} = x$.

**Step 1:**   Square both sides. $x + 6 = x^2$

**Step 2:**   Move everything to the side of $x^2$. $0 = x^2 - x - 6$

**Step 3:**   Here you may be able to see that $x^2 - x - 6$ conveniently factors to $(x-3)(x+2)$.
Solving $(x-3)(x+2) = 0$, we get $x = 3$ or $-2$.

**Example 10:**   Solve for $x$. $x = \sqrt{\dfrac{9x+54}{3}}$

**Step 1:**   Square both sides. $x^2 = \dfrac{9x+54}{3}$

**Step 2:**   Multiply both sides by 3. $3x^2 = 9x + 54$

**Step 3:**   Subtract $9x + 54$ from each side. $3x^2 - 9x - 54 = 0$

**Step 4:**   The expression $3x^2 - 9x - 54$ can be factored into $(3x+9)(x-6)$ because $54 = 9 \times -6$ and $9 + (3 \times -6) = -9$.
Solving $(3x+9)(x-6) = 0$, we find that $x = -3$ or $x = 6$.

**Solve for $x$.**

1. $x = \sqrt{18 - 7x}$

2. $x = \sqrt{\dfrac{7x-5}{2}}$

3. $x = \sqrt{\dfrac{5-9x}{2}}$

4. $x = \sqrt{\dfrac{-11x-6}{4}}$

5. $x = \sqrt{\dfrac{-6x-1}{5}}$

6. $\sqrt{-2x-1} = x$

7. $\sqrt{\dfrac{-8x-3}{5}} = x$

8. $x = \sqrt{\dfrac{-x}{2}}$

9. $x = \sqrt{\dfrac{5x+4}{6}}$

10. $x\sqrt{2} = \sqrt{-9x-10}$

11. $x = \sqrt{\dfrac{-25x-12}{12}}$

12. $x = \sqrt{\dfrac{3x}{2}}$

# Chapter 7 Review

**Factor and solve each of the following quadratic equations.**

1. $16b^2 - 25 = 0$

2. $a^2 - a - 30 = 0$

3. $x^2 - x = 6$

4. $100x^2 - 49 = 0$

5. $81y^2 = 9$

6. $y^2 = 21 - 4y$

7. $y^2 - 7y + 8 = 16$

8. $6x^2 + x - 2 = 0$

9. $3y^2 + y - 2 = 0$

10. $b^2 + 2b - 8 = 0$

11. $4x^2 + 19x - 5 = 0$

12. $8x^2 = 6x + 2$

13. $2y^2 - 6y - 20 = 0$

14. $-6x^2 + 7x - 2 = 0$

15. $y^2 + 3y - 18 = 0$

**Using the quadratic formula, find both solutions for the variable.**

16. $x^2 + 10x - 11 = 0$

17. $y^2 - 14y + 40 = 0$

18. $b^2 + 9b + 18 = 0$

19. $y^2 - 12y - 13 = 0$

20. $a^2 - 8a - 48 = 0$

21. $x^2 + 2x - 63 = 0$

**Use the following information for questions 22–24.**

Branden is tossing a ball in the air. He knows the height of the ball is represented by $s = -12t^2 + 50t$, where $s$ is the height and $t$ is the time.

22. How high (in feet) will the ball be after 4 seconds?

23. How high (in feet) will the ball be after 1.5 seconds?

24. How long (in seconds) was the ball in the air before it came back down to the ground?

**Solve for $x$.**

25. $x = \sqrt{\dfrac{3 - 4x}{4}}$

26. $x = \sqrt{\dfrac{7x}{2}}$

27. $x = \sqrt{\dfrac{-9x - 9}{2}}$

28. $x = \sqrt{\dfrac{3x - 1}{2}}$

## Chapter 7 Test

1. Solve: $4y^2 - 9y = -5$

   **A** $\left\{1, \frac{5}{4}\right\}$

   **B** $\left\{-\frac{3}{4}, -1\right\}$

   **C** $\left\{-1, \frac{4}{5}\right\}$

   **D** $\left\{\frac{5}{16}, 1\right\}$

2. Solve for $y$: $2y^2 + 13y + 15 = 0$

   **A** $\left\{\frac{3}{2}, \frac{5}{2}\right\}$

   **B** $\left\{\frac{2}{3}, \frac{2}{5}\right\}$

   **C** $\left\{-5, -\frac{3}{2}\right\}$

   **D** $\left\{5, -\frac{3}{2}\right\}$

3. Solve for $x$.

   $x^2 - 3x - 18 = 0$

   **A** $\{-6, 3\}$
   **B** $\{6, -3\}$
   **C** $\{-9, 2\}$
   **D** $\{9, -2\}$

4. What are the values of $x$ in the quadratic equation?

   $x^2 + 2x - 15 = x - 3$

   **A** $\{-4, 3\}$
   **B** $\{-3, 4\}$
   **C** $\{-3, 5\}$
   **D** Cannot be determined

5. Solve $6a^2 + 11a - 10 = 0$, using the quadratic formula.

   **A** $\left\{-\frac{2}{5}, \frac{3}{2}\right\}$

   **B** $\left\{\frac{2}{5}, \frac{2}{3}\right\}$

   **C** $\left\{-\frac{5}{2}, \frac{2}{3}\right\}$

   **D** $\left\{\frac{5}{2}, \frac{2}{3}\right\}$

6. Solve the equation $(x + 9)^2 = 49$

   **A** $x = -9, 9$
   **B** $x = -9, 7$
   **C** $x = -16, -2$
   **D** $x = -7, 7$

7. Solve the equation $c^2 + 8c - 9 = 0$ by completing the square.

   **A** $c = \{1, -9\}$
   **B** $c = \{-1, 9\}$
   **C** $c = \{3, 3\}$
   **D** $c = \{-3, -3\}$

8. Using the quadratic formula, solve the following equation:

   $3x^2 = 9x$

   **A** $x = \{0, 1\}$
   **B** $x = \{3, 1\}$
   **C** $x = \{0, 3\}$
   **D** $x = \{3, -3\}$

9. Eric and Jansen are throwing a baseball back and forth to each other. The height of the ball is represented by the equation $s = -8t^2 + 32t$. What is the highest point the ball will go?

   **A** 16 feet
   **B** 32 feet
   **C** 64 feet
   **D** 80 feet

10. Solve: $x = \sqrt{\dfrac{10 - 8x}{2}}$

    **A** $\{0, 6\}$
    **B** $\{-5, 10\}$
    **C** $\{4, -2\}$
    **D** $\{-5, 1\}$

# Chapter 8
# Graphing and Writing Equations and Inequalities

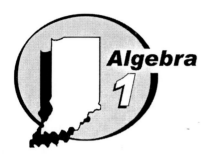

This chapter covers the following IN Algebra I standards:

| Standard 4: | Graphing Linear Equations and Inequalities | A1.4.1 |
| --- | --- | --- |
| | | A1.4.2 |
| | | A1.4.3 |
| | | A1.4.4 |
| | | A1.4.6 |

## 8.1    Graphing Linear Equations

In addition to graphing ordered pairs, the Cartesian plane can be used to graph the solution set for an equation. Any equation with two variables that are both to the first power is called a **linear equation.** The graph of a linear equation will always be a straight line.

**Example 1:**    Graph the solution set for $x + y = 7$.

**Step 1:**    Make a list of some pairs of numbers that will work in the equation.

$$\begin{array}{ll} \underline{x + y = 7} & \\ 4 + 3 = 7 & (4, 3) \\ -1 + 8 = 7 & (-1, 8) \\ 5 + 2 = 7 & (5, 2) \\ 0 + 7 = 7 & 0, 7 \end{array} \Bigg\} \text{ ordered pair solutions}$$

**Step 2:**    Plot these points on a Cartesian plane.

**Step 3:**    By passing a line through these points, we graph the solution set for $x + y = 7$. This means that every point on the line is a solution to the equation $x + y = 7$. For example, $(1, 6)$ is a solution, so the line passes through the point $(1, 6)$.

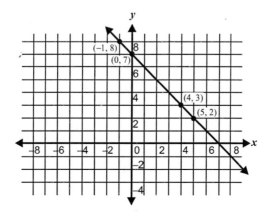

**Make a table of solutions for each linear equation below. Then plot the ordered pair solutions on graph paper. Draw a line through the points. (If one of the points does not line up, you have made a mistake.)**

1. $x + y = 6$

3. $y = x - 2$

5. $x - 5 = y$

2. $y = x + 1$

4. $x + 2 = y$

6. $x - y = 0$

**Example 2:**    Graph the equation $y = 2x - 5$.

**Step 1:**    This equation has 2 variables, both to the first power, so we know the graph will be a straight line. Substitute some numbers for $x$ or $y$ to find pairs of numbers that satisfy the equation. For the above equation, it will be easier to substitute values of $x$ in order to find the corresponding value for $y$. Record the values for $x$ and $y$ in a table.

| $x$ | $y$ |
|---|---|
| 0 | $-5$ |
| 1 | $-3$ |
| 2 | $-1$ |
| 3 | 1 |

If $x$ is 0, $y$ would be $-5$

If $x$ is 1, $y$ would be $-3$

If $x$ is 2, $y$ would be $-1$

If $x$ is 3, $y$ would be 1

**Step 2:**    Graph the ordered pairs, and draw a line through the points.

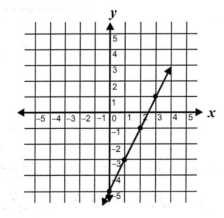

**Find pairs of numbers that satisfy the equations below, and graph the line on graph paper.**

1. $y = -2x + 2$

4. $y = x + 1$

7. $x = 4y - 3$

2. $2x - 2 = y$

5. $4x - 2 = y$

8. $2x = 3y + 1$

3. $-x + 3 = y$

6. $y = 3x - 3$

9. $x + 2y = 4$

## 8.2   Graphing Horizontal and Vertical Lines

The graph of some equations is a horizontal or a vertical line.

**Example 3:**   $y = 3$

**Step 1:**   Make a list of ordered pairs that satisfy the equation $y = 3$.

| $x$ | $y$ |
|---|---|
| 0 | 3 |
| 1 | 3 |
| 2 | 3 |
| 3 | 3 |

} No matter what value of $x$ you choose, $y$ is always 3.

**Step 2:**   Plot these points on an Cartesian plane, and draw a line through the points.

The graph is a horizontal line.

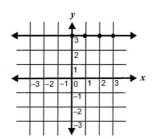

**Example 4:**   $2x + 3 = 0$

**Step 1:**   For these equations with only one variable, find what $x$ equals first.
$2x + 3 = 0$
$2x = -3$
$x = -\frac{3}{2}$

Just like Example 3, find ordered pairs that satisfy the equation, plot the points, and graph the line.

| $x$ | $y$ |
|---|---|
| $-\frac{3}{2}$ | 0 |
| $-\frac{3}{2}$ | 1 |
| $-\frac{3}{2}$ | 2 |
| $-\frac{3}{2}$ | 3 |

} No matter which value of $y$ you choose, the value of $x$ does not change.

The graph is a vertical line.

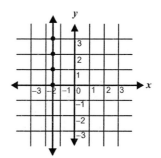

**Find pairs of numbers that satisfy the equations below, and graph the line on graph paper.**

1. $2y + 2 = 0$

2. $x = -4$

3. $3x = 3$

4. $y = 5$

5. $4x - 2 = 0$

6. $2x - 6 = 0$

7. $4y = 1$

8. $5x + 10 = 0$

9. $3y + 12 = 0$

10. $x + 1 = 0$

11. $2y - 8 = 0$

12. $3x = -9$

13. $x = -2$

14. $6y - 2 = 0$

15. $5x - 5 = 0$

## 8.3    Finding the Intercepts of a Line

The $x$-intercept is the point where the graph of a line crosses the $x$-axis. The $y$-intercept is the point where the graph of a line crosses the $y$-axis.

**To find the $x$-intercept, set $y = 0$**
**To find the $y$-intercept, set $x = 0$**

**Example 5:**    Find the $x$- and $y$-intercepts of the line $6x + 2y = 18$

**Step 1:**    To find the $x$-intercept, set $y = 0$.

$$
\begin{aligned}
6x + 2(0) &= 18 \\
6x &= 18 \\
\frac{6}{6} &\quad \frac{18}{6} \\
x &= 3
\end{aligned}
$$

The $x$-intercept is at the point $(3, 0)$.

**Step 2:**    To find the $y$-intercept, set $x = 0$.

$$
\begin{aligned}
6(0) + 2y &= 18 \\
2y &= 18 \\
\frac{2}{2} &\quad \frac{18}{2} \\
y &= 9
\end{aligned}
$$

The $y$-intercept is at the point $(0, 9)$.

You can now use the two intercepts to graph the line.

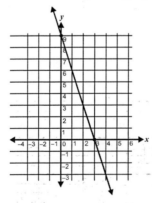

**For each of the following equations, find both the $x$ and the $y$ intercepts of the line. For extra practice, draw each of the lines on graph paper.**

1. $8x - 2y = 8$
2. $4x + 8y = 16$
3. $3x + 3y = 9$
4. $x - 2y = -5$
5. $8x + 4y = 32$

6. $3x - 4y = 12$
7. $-3x - 3y = 6$
8. $-6x + 2y = 18$
9. $4x - 2y = -4$
10. $-5x - 3y = 15$

11. $3x - 6y = -12$
12. $6x + 3y = 9$
13. $-2x - 6y = 18$
14. $2x + 3y = -6$
15. $-3x + 8y = 12$

## 8.4 Understanding Slope

The slope of a line refers to how steep a line is. Slope is also defined as the rate of change. When we graph a line using ordered pairs, we can easily determine the slope. Slope is often represented by the letter $m$.

The formula for slope of a line is: $m = \dfrac{y_2 - y_1}{x_2 - x_1}$ or $\dfrac{\text{rise}}{\text{run}}$

**Example 6:** What is the slope of the following line that passes through the ordered pairs $(-4, -3)$ and $(1, 3)$?

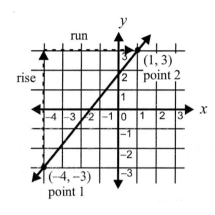

$y_2$ is 3, the $y$-coordinate of point 2.

$y_1$ is $-3$, the $y$-coordinate of point 1.

$x_2$ is 1, the $x$-coordinate of point 2.

$x_1$ is $-4$, the $x$-coordinate of point 1.

Use the formula for slope given above:

$$m = \frac{3 - (-3)}{1 - (-4)} = \frac{6}{5}$$

The slope is $\frac{6}{5}$. This shows us that we can go up 6 (rise) and over 5 to the right (run) to find another point on the line.

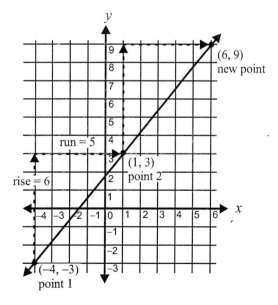

**Example 7:**   Find the slope of a line through the points $(-2, 3)$ and $(1, -2)$. It doesn't matter which pair we choose for point 1 and point 2. The answer is the same.

Let point 1 be $(-2, 3)$
Let point 2 be $(1, -2)$

$$\text{slope} = \frac{(y_2 - y_1)}{(x_2 - x_1)} = \frac{-2 - 3}{1 - (-2)} = \frac{-5}{3}$$

When the slope is negative, the line will slant left. For this example, the line will go **down** 5 units and then over 3 units to the **right**.

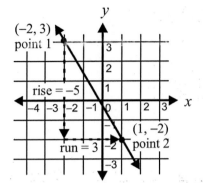

**Example 8:**   What is the slope of a line that passes through $(1, 1)$ and $(3, 1)$?

$$\text{slope} = \frac{1 - 1}{3 - 1} = \frac{0}{2} = 0$$

When $y_2 - y_1 = 0$, the slope will equal 0, and the line will be horizontal.

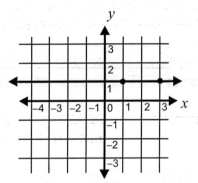

**Example 9:**   What is the slope of a line that passes through $(2, 1)$ and $(2, -3)$?

$$\text{slope} = \frac{-3 - 1}{2 - 2} = \frac{-4}{0} = \text{undefined}$$

When $x_2 - x_1 = 0$, the slope is undefined, and the line will be vertical.

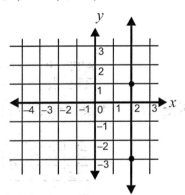

**The following lines summarize what we know about slope.**

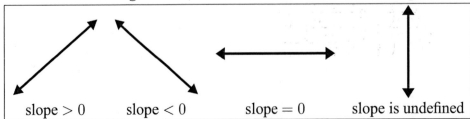

slope $> 0$          slope $< 0$          slope $= 0$          slope is undefined

**Find the slope of the line that goes through the following pairs of points. Then, using graph paper, graph the line through the two points, and label the rise and run. (See Examples 6–9).**

1. $(2, 3)$ $(4, 5)$
2. $(1, 3)$ $(2, 5)$
3. $(-1, 2)$ $(4, 1)$
4. $(1, -2)$ $(4, -2)$

5. $(3, 0)$ $(3, 4)$
6. $(3, 2)$ $(-1, 8)$
7. $(4, 3)$ $(2, 4)$
8. $(2, 2)$ $(1, 5)$

9. $(3, 4)$ $(1, 2)$
10. $(3, 2)$ $(3, 6)$
11. $(6, -2)$ $(3, -2)$
12. $(1, 2)$ $(3, 4)$

13. $(-2, 1)$ $(-4, 3)$
14. $(5, 2)$ $(4, -1)$
15. $(1, -3)$ $(-2, 4)$
16. $(2, -1)$ $(3, 5)$

## 8.5   Slope-Intercept Form of a Line

An equation that contains two variables, each to the first degree, is a **linear equation**. The graph for a linear equation is a straight line. To put a linear equation in slope-intercept form, solve the equation for $y$. This form of the equation shows the slope and the $y$-intercept. Slope-intercept form follows the pattern of $y = mx + b$. The "$m$" represents slope, and the "$b$" represents the $y$-intercept. The $y$-intercept is the point at which the line crosses the $y$-axis.

When the slope of a line is not 0, the graph of the equation shows a **direct variation** between $y$ and $x$. When $y$ increases, $x$ increases in a certain proportion. The proportion stays constant. The constant is called the **slope** of the line.

**Example 10:**   Put the equation $2x + 3y = 15$ in slope-intercept form. What is the slope of the line? What is the $y$-intercept? Graph the line.

**Step 1:**   Solve for $y$:

$$\begin{array}{rcr} 2x + 3y &=& 15 \\ -2x & & -2x \\ \hline \dfrac{3y}{3} &=& -\dfrac{2x}{3} + \dfrac{15}{3} \end{array}$$

slope-intercept form:   $y = -\frac{2}{3}x + 5$
The slope is $-\frac{2}{3}$ and the $y$-intercept is 5.

**Step 2:**   Knowing the slope and the $y$-intercept, we can graph the line.

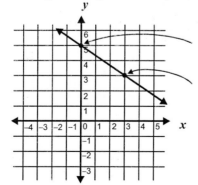

The $y$-intercept is 5, so the line passes through the point $(0, 5)$ on the $y$-axis.

The slope is $-\frac{2}{3}$, so go down 2 and over 3 to get a second point.

**Put each of the following equations in slope-intercept form by solving for $y$. On your graph paper, graph the line using the slope and $y$-intercept.**

1. $4x - 5y = 5$

2. $2x + 4y = 16$

3. $3x - 2y = 10$

4. $x + 3y = -12$

5. $6x + 2y = 0$

6. $8x - 5y = 10$

7. $-2x + y = 4$

8. $-4x + 3y = 12$

9. $-6x + 2y = 12$

10. $x - 5y = 5$

11. $3x - 2y = -6$

12. $3x + 4y = 2$

13. $-x = 2 + 4y$

14. $2x = 4y - 2$

15. $6x - 3y = 9$

16. $4x + 2y = 8$

17. $6x - y = 4$

18. $-2x - 4y = 8$

19. $5x + 4y = 16$

20. $6 = 2y - 3x$

## 8.6   Graphing a Line Knowing a Point and Slope

If you are given a point of a line and the slope of a line, the line can be graphed.

**Example 11:**   Given that line $l$ has a slope of $\frac{4}{3}$ and contains the point $(2, -1)$, graph the line.

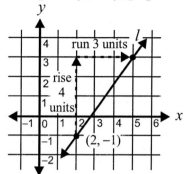

Plot and label the point $(2, -1)$ on a Cartesian plane.

The slope, $m$, is $\frac{4}{3}$, so the rise is 4, and the run is 3. From the point $(2, -1)$, count 4 units up and 3 units to the right.

Draw the line through the two points.

**Example 12:**   Given a line that has a slope of $-\frac{1}{4}$ and passes through the point $(-3, 2)$, graph the line.

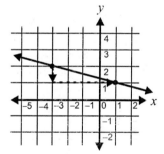

Plot the point $(-3, 2)$.

Since the slope is negative, go **down** 1 unit and over 4 units to get a second point.

Graph the line through the two points.

**Graph a line on your own graph paper for each of the following problems. First, plot the point. Then use the slope to find a second point. Draw the line formed from the point and the slope.**

1. $(2, -2)$, $m = \frac{3}{4}$

2. $(3, -4)$, $m = \frac{1}{2}$

3. $(1, 3)$, $m = -\frac{1}{3}$

4. $(2, -4)$, $m = 1$

5. $(3, 0)$, $m = -\frac{1}{2}$

6. $(-2, 1)$, $m = \frac{4}{3}$

7. $(-4, -2)$, $m = \frac{1}{2}$

8. $(1, -4)$, $m = \frac{3}{4}$

9. $(2, -1)$, $m = -\frac{1}{2}$

10. $(5, -2)$, $m = \frac{1}{4}$

11. $(-2, -3)$, $m = \frac{2}{3}$

12. $(4, -1)$, $m = -\frac{1}{3}$

13. $(-1, 5)$, $m = \frac{2}{5}$

14. $(-2, 3)$, $m = \frac{3}{4}$

15. $(4, 4)$, $m = -\frac{1}{2}$

16. $(3, -3)$, $m = -\frac{3}{4}$

17. $(-2, 5)$, $m = \frac{1}{3}$

18. $(-2, -3)$, $m = -\frac{3}{4}$

19. $(4, -3)$, $m = \frac{2}{3}$

20. $(1, 4)$, $m = -\frac{1}{2}$

## 8.7   Finding the Equation of a Line Using Two Points or a Point and Slope

If you can find the slope of a line and know the coordinates of one point, you can write the equation for the line. You know the formula for the slope of a line is:

$$m = \frac{y_2 - y_1}{x_2 - x_1} \text{ or } \frac{y_2 - y_1}{x_2 - x_1} = m$$

Using algebra, you can see that if you multiply both sides of the equation by $x_2 - x_1$, you get:

$$y - y_1 = m(x - x_1) \longleftarrow \text{point-slope form of an equation}$$

**Example 13:**   Write the equation of the line passing through the points $(-2, 3)$ and $(1, 5)$.

**Step 1:**   First, find the slope of the line using the two points given.
$$m = \frac{y_2 - y_1}{x_2 - x_1} = \frac{5 - 3}{1 - (-2)} = \frac{2}{3}$$

**Step 2:**   Pick one of the two points to use in the point-slope equation. For point $(-2, 3)$, we know $x_1 = -2$ and $y_1 = 3$, and we know $m = \frac{2}{3}$. Substitute these values into the point-slope form of the equation.
$$y - y_1 = m(x - x_1)$$
$$y - 3 = \frac{2}{3}[x - (-2)]$$
$$y - 3 = \frac{2}{3}x + \frac{4}{3}$$
$$y = \frac{2}{3}x + \frac{13}{3}$$

**Use the point-slope formula to write an equation for each of the following lines.**

1. $(1, -2)$, $m = 2$

2. $(-3, 3)$, $m = \frac{1}{3}$

3. $(4, 2)$, $m = \frac{1}{4}$

4. $(5, 0)$, $m = 1$

5. $(3, -4)$, $m = \frac{1}{2}$

6. $(-1, -4)$ $(2, -1)$

7. $(2, 1)$ $(-1, -3)$

8. $(-2, 5)$ $(-4, 3)$

9. $(-4, 3)$ $(2, -1)$

10. $(3, 1)$ $(5, 5)$

11. $(-3, 1)$, $m = 2$

12. $(-1, 2)$, $m = \frac{4}{3}$

13. $(2, -5)$, $m = -2$

14. $(-1, 3)$, $m = \frac{1}{3}$

15. $(0, -2)$, $m = -\frac{3}{2}$

## 8.8   Graphing Inequalities

In the previous section, you would graph the equation $x = 3$ as:

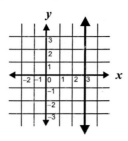

In this section, we graph inequalities such as $x > 3$ (read $x$ is greater than 3). To show this, we use a broken line since the points on the line $x = 3$ are not included in the solution. We shade all points greater than 3.

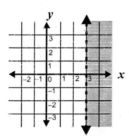

When we graph $x \geq 3$ (read $x$ is greater than or equal to 3), we use a solid line because the points on the line $x = 3$ are included in the graph.

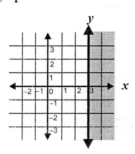

**Graph the following inequalities on your own graph paper.**

1. $y < 2$
2. $x \geq 4$
3. $y \geq 1$
4. $x < -1$
5. $y \geq -2$
6. $x \leq -4$

7. $x > -3$
8. $y \leq 3$
9. $x \leq 5$
10. $y > -5$
11. $x \geq 3$
12. $y < -1$

13. $x \leq 0$
14. $y > -1$
15. $y \leq 4$
16. $x \geq 0$
17. $y \geq 3$
18. $x < 4$

19. $x \leq -2$
20. $y < -2$
21. $y \geq -4$
22. $x \geq -1$
23. $y \leq 5$
24. $x < -3$

**Example 14:**     Graph $x + y \geq 3$.

**Step 1:**     First, we graph $x + y \geq 3$ by changing the inequality to an equality. Think of ordered pairs that will satisfy the equation $x + y = 3$. Then, plot the points, and draw the line. As shown below, this line divides the Cartesian plane into 2 half-planes, $x + y \geq 3$ and $x + y \leq 3$. One half-plane is above the line, and the other is below the line.

| $x$ | $y$ |
|-----|-----|
| 2 | 1 |
| 0 | 3 |
| 3 | 0 |
| 4 | −1 |

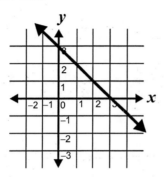

**Step 2:**     To determine which side of the line to shade, first choose a test point. If the point you choose makes the inequality true, then the point is on the side you shade. If the point you choose does not make the inequality true, then shade the side that does not contain the test point.

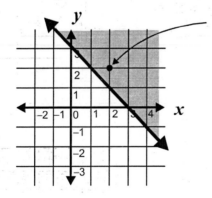

For our test point, let's choose $(2, 2)$. Substitute $(2, 2)$ into the inequality.

$x + y \geq 3$
$2 + 2 \geq 3$

$4 \geq 3$ is true, so shade the side that includes this point.

Use a solid line because of the $\geq$ sign.

**Graph the following inequalities on your own graph paper.**

1.  $x + y \leq 4$
2.  $x + y \geq 3$
3.  $x \geq 5 - y$
4.  $x \leq 1 + y$

5.  $x - y \geq -2$
6.  $x < y + 4$
7.  $x + y < -1$
8.  $x - y \leq 0$

9.  $x \geq y + 2$
10.  $x < -y + 1$
11.  $-x + y > 1$
12.  $-x - y < -2$

For more complex inequalities, it is easier to graph by first changing the inequality to an equality and then put the equation in slope-intercept form.

**Example 15:**   Graph the inequality $2x + 4y \leq 8$.

**Step 1:**   Change the inequality to an equality.

$$2x + 4y = 8$$

**Step 2:**   Put the equation in slope-intercept form by solving the equation for $y$.

$$2x + 4y = 8$$
$$2x - 2x + 4y = -2x + 8 \qquad \text{Subtract } 2x \text{ from both sides of the equation.}$$
$$4y = -2x + 8 \qquad \text{Simplify.}$$
$$\frac{4y}{4} = \frac{-2x + 8}{4} \qquad \text{Divide both sides by 4.}$$
$$y = \frac{-2x}{4} + \frac{8}{4} \qquad \text{Find the lowest terms of the fractions.}$$
$$y = -\tfrac{1}{2}x + 2$$

**Step 3:**   Graph the line. If the inequality is $<$ or $>$, use a dotted line. If the inequality is $\leq$ or $\geq$, use a solid line. For this example, we should use a solid line.

**Step 4:**   Determine which side of the line to shade. Pick a point such as $(0,0)$ to see if it is true in the inequality.

$2x + 4y \leq 8$, so substitute $(0,0)$.
Is $0 + 0 \leq 8$? Yes, $0 \leq 8$, so shade the side of the line that includes the point $(0,0)$.

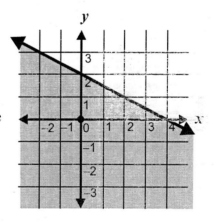

**Graph the following inequalities on your own graph paper.**

1.  $2x + y \geq 1$

2.  $3x - y \leq 3$

3.  $x + 3y > 12$

4.  $4x - 3y < 12$

5.  $y \geq 3x + 1$

6.  $x - 2y > -2$

7.  $x \leq y + 4$

8.  $x + y < -1$

9.  $-4y \geq 2x + 1$

10.  $x \leq 4y - 2$

11.  $3x - y \geq 4$

12.  $y \geq 2x - 5$

## Chapter 8 Review

1. Graph the solution set for the linear equation: $x - 3 = y$.

2. Graph the equation $2x - 4 = 0$.

3. What is the slope of the line that passes through the points $(5, 3)$ and $(6, 1)$?

4. What is the slope of the line that passes through the points $(-1, 4)$ and $(-6, -2)$?

5. What is the $x$-intercept for the following equation? $6x - y = 30$

6. What is the $y$-intercept for the following equation? $4x + 2y = 28$

7. Graph the equation $3y = 9$.

8. Write the following equation in slope-intercept form.
$$3x = -2y + 4$$

9. What is the slope of the line $y = -\frac{1}{2}x + 3$?

10. What is the $x$-intercept of the line $y = 5x + 6$?

11. What is the $y$-intercept of the line $y - \frac{2}{3}x + 3 = 0$?

12. Graph the line which has a slope of $-2$ and a $y$-intercept of $-3$.

13. Find the equation of the line which contains the point $(0, 2)$ and has a slope of $\frac{3}{4}$.

**Graph the following inequalities on a Cartesian plane using your graph paper.**

14. $x \geq 4$

15. $x \leq -2$

16. $5y > -10x + 5$

17. $y \leq 2$

18. $2x + y < 5$

19. $y - 2x \leq 3$

20. $y \geq x + 2$

21. $3 + y > x$

# Chapter 8 Test

1. Which of the following is not a solution of $3x = 5y - 1$?

   **A** $(3, 2)$
   **B** $(7, 4)$
   **C** $\left(-\frac{1}{3}, 0\right)$
   **D** $(-2, -1)$

2. $(-2, 1)$ is a solution for which of the following equations?

   **A** $y + 2x = 4$
   **B** $-2x - y = 5$
   **C** $x + 2y = -4$
   **D** $2x - y = -5$

3. Which is the graph of $x - 3y = 6$?

   **A**
   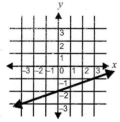

   **B**
   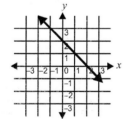

   **C**
   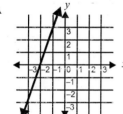

   **D**
   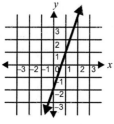

4. Which of the following points does **not** lie on the line $y = 3x - 2$?

   **A** $(0, -2)$
   **B** $(1, 1)$
   **C** $(-1, 5)$
   **D** $(2, 4)$

5. Which of the following is the graph of the equation $y = x - 3$?

   **A**
   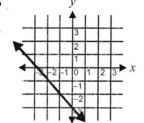

   **B**
   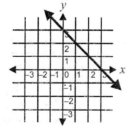

   **C**
   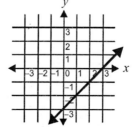

   **D**
   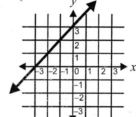

6. What is the $x$-intercept of the following linear equation? $3x + 4y = 12$

   **A** $(0, 3)$
   **B** $(3, 0)$
   **C** $(0, 4)$
   **D** $(4, 0)$

7. Which of the following equations is represented by the graph?

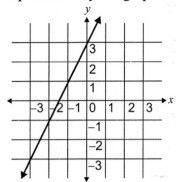

   **A** $y = -3x + 3$

   **B** $y = -\frac{1}{3}x + 3$

   **C** $y = 3x - 3$

   **D** $y = 2x + 3$

8. What is the equation of the line that includes the point $(4, -3)$ and has a slope of $-2$?

   **A** $y = -2x - 5$
   **B** $y = -2x - 2$
   **C** $y = -2x + 5$
   **D** $y = 2x - 5$

9. What is the $x$-intercept and $y$-intercept for the equation $x + 2y = 6$?

   **A** $x$-intercept $= (0, 6)$
   $y$-intercept $= (3, 0)$
   **B** $x$-intercept $= (4, 1)$
   $y$-intercept $= (2, 2)$
   **C** $x$-intercept $= (0, 6)$
   $y$-intercept $= (0, 3)$
   **D** $x$-intercept $= (6, 0)$
   $y$-intercept $= (0, 3)$

10. Put the following equation in slope-intercept form.

    $$2x - 3y = 6$$

    **A** $y = \frac{2}{3}x - 2$
    **B** $y = 2x - 2$
    **C** $y = -\frac{2}{3}x + 2$
    **D** $y = 2x + 2$

11. Which of the following is a graph of the inequality $-y \geq 2$?

    **A**

    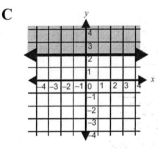

    **B**

    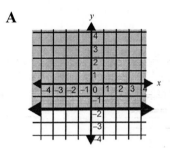

    **C**

    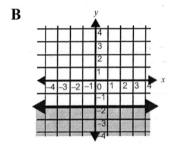

    **D**

    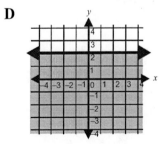

12. Which of the following graphs shows a line with a slope of 0 that passes through the point $(3, 2)$?

**A**

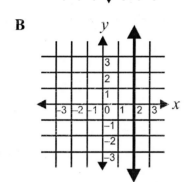

**B**

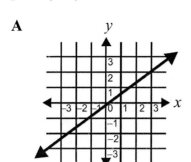

**C**

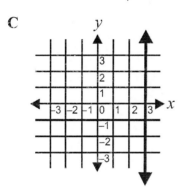

**D**

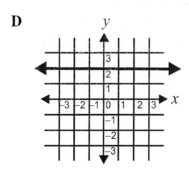

13. Which of the following is a graph of the inequality $y \leq x - 3$?

**A**

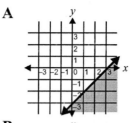

**B**

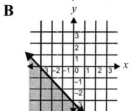

**C**

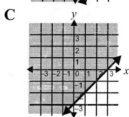

**D**

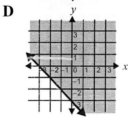

14. Look at the graphs below. Which of the following statements is false?

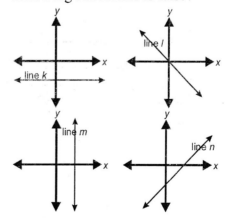

**A** The slope of line $k$ is undefined.
**B** The slope of line $l$ is negative.
**C** The slope of line $m$ is undefined.
**D** The slope of line $n$ is positive.

# Chapter 9
# Applications of Graphs

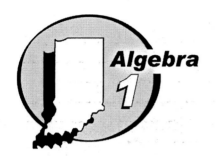

This chapter covers the following IN Algebra I standards:

| Standard 3: | Relations and Functions | A1.3.1 |
|---|---|---|
| Standard 4: | Graphing Linear Equations | A1.4.4 |
| | and Inequalities | A1.4.5 |
| Standard 8: | Quadratic, Cubic, and Radical | A1.8.1 |
| | Equations | A1.8.9 |

## 9.1    Changing the Slope or $Y$-Intercept of a Line

When the slope and/or the $y$-intercept of a linear equation changes, the graph of the line will also change.

**Example 1:**      Consider line $l$ shown in Figure 1 at right. What happens to the graph of the line if the slope is changed to $\frac{4}{5}$?

Determine the $y$-intercept of the line. For line $l$, it can easily be seen from the graph that the $y$-intercept is at the point $(0, -1)$.

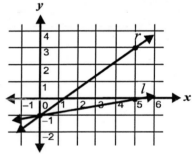

Figure 1

Find the slope of the line using two points that the line goes through: $(0, -1)$ and $(5, 0)$.

$$m = \frac{y_2 - y_1}{x_2 - x_2} = \frac{0 - (-1)}{5 - 0} = \frac{1}{5}$$

Write the equation of line $l$ in slope-intercept form:
$$y = mx + b \quad \Longrightarrow \quad y = \tfrac{1}{5}x - 1$$

Rewrite the equation of the line using a slope of $\frac{4}{5}$, and then graph the line. The equation of the new line is $y = \frac{4}{5}x - 1$.

The graph of the new line is labeled line $r$ and is shown in Figure 1. A line with a slope of $\frac{4}{5}$ is steeper than a line with a slope of $\frac{1}{5}$.

**Note: The greater the numerator, or "rise," of the slope, the steeper the line will be. The greater the denominator, or "run," of the slope, the flatter the line will be.**

**Consider the line ($l$) shown on each of the following graphs, and write the equation of the line in the space provided. Then, on the same graph, graph the line ($r$) for which the equation is given. Write how the slope and $y$-intercept of line $l$ compare to the slope and $y$-intercept of line $r$ for each graph.**

1.

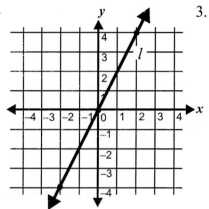

line $l$: _____

line $r$: ___ $y = -2x$ ___

slopes: _____

$y$-intercepts: _____

3.

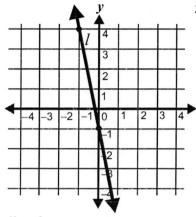

line $l$: _____

line $r$: ___ $y = -3x - 1$ ___

slopes: _____

$y$-intercepts: _____

5.

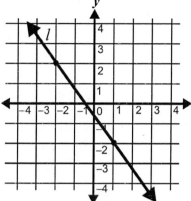

line $l$: _____

line $r$: ___ $y = \frac{1}{4}x - 2$ ___

slopes: _____

$y$-intercepts: _____

2.

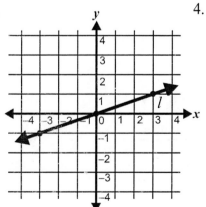

line $l$: _____

line $r$: ___ $y = \frac{1}{3}x + 2$ ___

slopes: _____

$y$-intercepts: _____

4.

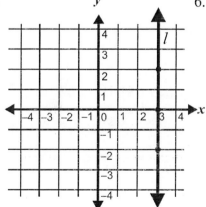

line $l$: _____

line $r$: ___ $y = -3$ ___

slopes: _____

$y$-intercepts: _____

6.

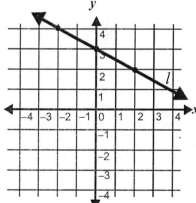

line $l$: _____

line $r$: ___ $y = -\frac{1}{2}x - 3$ ___

slopes: _____

$y$-intercepts: _____

## 9.2    Equations of Parallel Lines

If two linear equations have the same slope but different $y$-intercepts, they are **parallel** lines. Parallel lines never touch each other, so they have no points in common.

**Example 2:**    Consider line $l$ shown in Figure 2 at right. The equation of the line is $y = -\frac{1}{2}x + 3$. What happens to the graph of the line if the $y$-intercept is changed to $-1$?

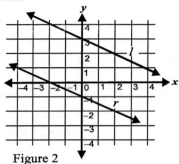

Figure 2

Rewrite the equation of the line replacing the $y$-intercept with $-1$. The equation of the new line is $y = -\frac{1}{2}x - 1$.

Graph the new line. Line $r$ in Figure 2 is the graph of the equation $y = -\frac{1}{2}x - 1$. Since both lines $l$ and $r$ have the same slope, they are parallel. Line $r$, with a $y$-intercept of $-1$, sits below line $l$, with a $y$-intercept of 3.

**Put each pair of the following equations in slope-intercept form.  Write P if the lines are parallel and NP if the lines are not parallel.**

1.  $y = x + 1$        _____
    $2y - 2x = 6$

2.  $2x + y = 6$        _____
    $2x = 8 - y$

3.  $x + 5y = 0$        _____
    $5y + 5 = x$

4.  $y = 3 - \frac{1}{3}x$        _____
    $3y + x = -6$

5.  $x = 2y$        _____
    $-x = -2y + 14$

6.  $y = x + 2$        _____
    $-y = x + 4$

7.  $y = 4 - \frac{1}{4}x$        _____
    $3x + 4y = 4$

8.  $x + y = 5$        _____
    $5 - y = 2x$

9.  $x - 4y = 0$        _____
    $4y = x - 8$

## 9.3    Equations of Perpendicular Lines

Now that we know how to calculate the slope of lines using two points, we are going to learn how to calculate the slope of a line perpendicular to a given line, then find the equation of that perpendicular line. To find the slope of a line perpendicular to any given line, take the slope of the first line, $m$:

1. multiply the slope by $-1$

2. invert (or flip over) the slope

You now have the slope of a perpendicular line. Writing the equation for a line perpendicular to another line involves three steps:

1. Find the slope of the perpendicular line.

2. Choose one point on the first line.

3. Use the point-slope form to write the equation.

**Example 3:**      The solid line on the graph below has a slope of $\frac{2}{3}$. Write the equation of a line perpendicular to the solid line.

**Step 1:**      Find the slope of the solid line. Multiply the slope by $-1$ and then find the inverse (flip it over).

$$\frac{2}{3} \times -1 = -\frac{2}{3} \curvearrowright -\frac{3}{2}$$

The slope of the perpendicular line, shown as a dotted line on the graph below, is $-\frac{3}{2}$.

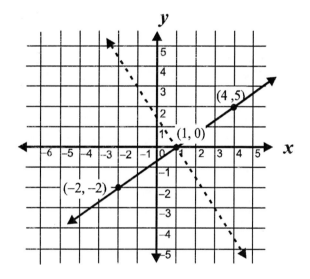

**Step 2:**      Choose one point on the first line. We will use $(1, 0)$ in this example. The point $(-2, -2)$ or $(4, 5)$ could also be used.

**Step 3:**      Use the point-slope formula, $(y - y_1) = m(x - x_1)$, to write the equation of the perpendicular line. Remember, we chose $(1, 0)$ as our point. So, $(y - 0) = -\frac{3}{2}(x - 1)$. Simplified, $y = -\frac{3}{2}x + \frac{3}{2}$.

127

## Solve the following problems involving perpendicular lines.

1. Find the slope of the line perpendicular to the solid line shown at right, and draw the perpendicular as a dotted line. Use the point $(-1, 0)$ on the solid line and the calculated slope to find the equation of the perpendicular line.

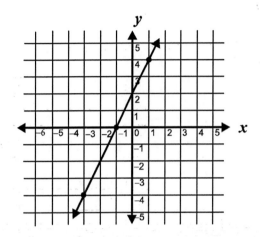

**Find the equation of the perpendicular line using the point and slope given and the formula** $(y - y_1) = m(x - x_1)$.

2. $(2, 1), 5$

3. $(3, 2), 2$

4. $(-2, 1), -3$

5. $(-4, 2), -\dfrac{1}{2}$

6. $(-1, 4), 1$

7. $(3, 3), \dfrac{2}{3}$

8. $(5, -1), -1$

9. $\left(\dfrac{1}{2}, \dfrac{3}{4}\right), 4$

10. $\left(\dfrac{2}{3}, \dfrac{3}{4}\right), -\dfrac{1}{6}$

11. $(7, -2), -\dfrac{1}{8}$

12. $(5, 0), \dfrac{4}{5}$

13. $(-3, -3), -\dfrac{7}{3}$

14. $\left(\dfrac{1}{4}, 4\right), \dfrac{1}{2}$

15. $(0, 6), -\dfrac{1}{9}$

## 9.4 Graphing Quadratic Equations

Equations that you may encounter on the IN Algebra 1 state exam may involve variables which are squared. The best way to find values for the $x$ and $y$ variables in an equation is to plug one number into $x$, and then find the corresponding value for $y$ just as you did at the beginning of this chapter. Then, plot the points and draw a line through the points.

**Example 4:**    Graph $y = x^2$.

**Step 1:**    Make a table and find several values for $x$ and $y$.

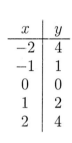

| $x$ | $y$ |
|----|----|
| $-2$ | 4 |
| $-1$ | 1 |
| 0 | 0 |
| 1 | 2 |
| 2 | 4 |

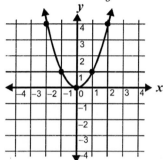

**Step 2:**    Plot the points, and draw a curve through the points. Notice the shape of the curve. This type of curve is called a **parabola**.

**Example 5:**    Graph the equation $y = -2x^2 + 4$.

**Step 1:**    Make a table and find several values for $x$ and $y$.

| $x$ | $y$ |
|----|----|
| $-2$ | $-4$ |
| $-1$ | 2 |
| 0 | 4 |
| 1 | 2 |
| 2 | $-4$ |

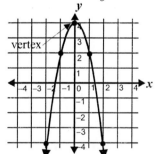

**Step 2:**    Plot the points, and draw a curve through the points.

**Note:** In the equation $y = ax^2 + c$, changing the value of $a$ will widen or narrow the parabola around the $y$-axis. If the value of $a$ is a negative number, the parabola will be reflected across the $x$-axis (the vertex will be at the top of the parabola instead of at the bottom). If $a = 0$, the graph will be a straight line, not a parabola. Changing the value of $c$ will move the vertex of the parabola from the origin to a different point on the $y$-axis.

**Graph the equations below on a Cartesian plane.**

1. $y = 2x^2$
2. $y = 3 - x^2$
3. $y = x^2 - 2$

4. $y = -2x^2$
5. $y = x^2 + 3$
6. $y = -3x^2 + 2$

7. $y = 3x^2 - 5$
8. $y = x^2 + 1$
9. $y = -x^2 - 6$

10. $y = -x^2$
11. $y = 2x^2 - 1$
12. $y = 2 - 2x^2$

## 9.5   Finding the Intercepts of a Quadratic Equation

A **quadratic equation** is an equation where either the $y$ or $x$ variable is squared. Finding the intercepts of a quadratic equation is similar to finding the intercepts of a line. In most cases, the variable $x$ is squared, which means there could be two $x$-intercepts. There could be one, two, or zero $x$-intercepts. The $x$-intercepts of a quadratic equation are called the **roots** of a quadratic equation.

**Example 6:**   Find the $x$-intercept(s) and the $y$-intercept(s) of the quadratic equation, $y = x^2 - 4$.

**Step 1:**   First, find the $y$-intercept. Since the $y$-intercept is the point where the graph crosses the $y$-axis, the value for $x$ at this point is zero. Because we know that $x = 0$, plug 0 in for $x$ in the equation and solve for $y$.
$$y = x^2 - 4$$
$$y = 0^2 - 4$$
$$y = 0 - 4$$
$$y = -4$$
Therefore, the $y$-intercept is $(0, -4)$.

**Step 2:**   Next, find the $x$-intercept(s). The $x$-intercept is the point, or points in this case, where the graph crosses the $x$-axis. In this case, we plug 0 in for $y$ because $y$ is always zero along the $x$-axis. Solve for $x$.
$$y = x^2 - 4$$
$$0 = x^2 - 4$$
$$0 + 4 = x^2 - 4 + 4$$
$$4 = x^2$$
$$\sqrt{4} = \sqrt{x^2}$$
$$\sqrt{4} = x$$
Thus, $x = \sqrt{4}$ and $\sqrt{4} = -2$ or 2, so $x = -2$ or $x = 2$.
The $x$-intercepts are $(-2, 0)$ and $(2, 0)$.

**Step 3:**   To verify that the intercepts are correct, graph the equation on the coordinate plane.

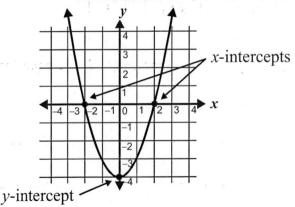

**Example 7:**   Find the $x$-intercept and the $y$-intercept of the quadratic equation, $y = x^2 + 2$.

**Step 1:**   First, find the $y$-intercept. Plug 0 in for $x$ in the equation and solve for $y$.
$y = x^2 + 2 = 0^2 + 2 = 0 + 2 = 2$
Therefore, the $y$-intercept is $(0, 2)$.

**Step 2:**   Next, find the $x$-intercept. Plug 0 in for $y$ and solve for $x$.
$y = x^2 + 2$
$0 = x^2 + 2$
$0 - 2 = x^2 + 2 - 2$
$-2 = x^2$
$\sqrt{-2} = \sqrt{x^2}$
$\sqrt{-2} = x$
You cannot take the square root of a negative number and get a real number as the answer, so there is no $x$-intercept for this quadratic equation.

**Step 3:**   To verify that the intercepts are correct, graph the equation on the coordinate plane. As you can see from the graph, the tails of the parabola are increasing in the positive $y$ direction, so they will never come back down, which means they will never cross the $x$-axis.

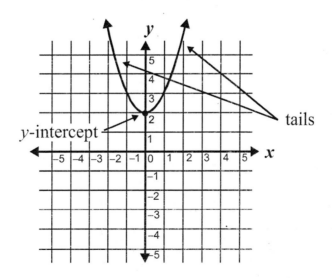

**Find the $x$-intercept and $y$-intercept of the following quadratic equations.**

1. $y = x^2$

2. $y = 2x^2 - 4$

3. $y = -x^2 + 2$

4. $y = x^2 + 6$

5. $y = -x^2 - 1$

6. $y = x^2 - 5$

7. $y = 4x^2 + 8$

8. $y = x^2 + 2x + 1$

9. $y = x^2 - 7x + 12$

## 9.6    Graphing Basic Functions

A graph is an image that shows the relationship between two or more variables. In this section, we will learn how to graph six basic functions.

**Example 8:**    $f(x) = x$

*The notation $f(x)$ is the same as $y$, it just means $f$ as a function of $x$*

The easiest way to begin graphing this is to draw a table of values. To create an accurate graph, you should choose at least 5 values for your table. These values are now your $(x, y)$ ordered pair and they are ready to be plotted.

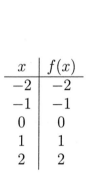

| $x$ | $f(x)$ |
|-----|--------|
| $-2$ | $-2$ |
| $-1$ | $-1$ |
| $0$ | $0$ |
| $1$ | $1$ |
| $2$ | $2$ |

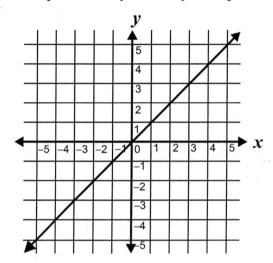

This is called the graph of a **linear function**.

**Example 9:**    $f(x) = x^2$

Exponential functions can be graphed the same way as linear functions.

| $x$ | $f(x)$ |
|-----|--------|
| $-2$ | $4$ |
| $-1$ | $1$ |
| $0$ | $0$ |
| $1$ | $1$ |
| $2$ | $4$ |

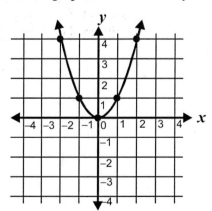

This is called the graph of a **quadratic function**.

**Example 10:**  $f(x) = x^3$

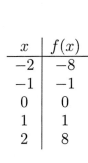

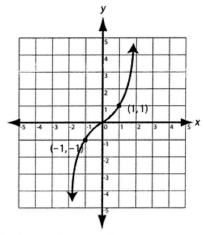

| $x$ | $f(x)$ |
|---|---|
| $-2$ | $-8$ |
| $-1$ | $-1$ |
| $0$ | $0$ |
| $1$ | $1$ |
| $2$ | $8$ |

This is called the graph of a **cubic function**.

**Example 11:**  $f(x) = \sqrt{x}$

An equation like this will be easier to graph if you choose values of $x$ that are perfect squares and because you can't take the square root of a negative, there is no need to select negative values.

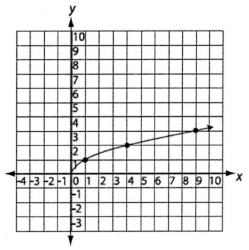

| $x$ | $f(x)$ |
|---|---|
| $0$ | $0$ |
| $1$ | $1$ |
| $4$ | $2$ |
| $9$ | $3$ |

This is called the graph of a **radical function**.

**Graph the following.**

1. $f(x) = 2x$

2. $f(x) = 3x^3$

3. $f(x) = \sqrt{2x}$

4. $f(x) = \frac{1}{3}x^2$

5. $f(x) = -4x^2$

6. $f(x) = -\sqrt{x}$

## 9.7   Writing an Equation From Data

Data is often written in a two-column format. If the increases or decreases in the ordered pairs are at a constant rate, then a linear equation for the data can be found.

**Example 12:**   Write an equation for the following set of data.

Dan set his car on cruise control and noted the distance he went every 5 minutes.

| Minutes in operation $(x)$ | Odometer reading $(y)$ |
|:---:|:---:|
| 5 | 28,490 miles |
| 10 | 28,494 miles |

**Step 1:**   Write two order pairs in the form (minutes, distance) for Dan's driving, $(5, 28490)$ and $(10, 28494)$, and find the slope.
$$m = \frac{28494 - 28490}{10 - 5} = \frac{4}{5}$$

**Step 2:**   Use the ordered pairs to write the equation in the form $y = mx + b$. Place the slope, $m$, that you found and one of the pairs of points as $x_1$ and $y_1$ in the following formula, $y - y_1 = m(x - x_1)$.

$y - 28490 = \frac{4}{5}(x - 5)$
$y - 28490 = \frac{4}{5}x - 4$
$y - 28490 + 28490 = \frac{4}{5}x - 4 + 28490$
$y + 0 = \frac{4}{5}x + 28486$
$y = \frac{4}{5}x + 28486$

**Write an equation for each of the following sets of data, assuming the relationship is linear.**

1.

**Doug's Doughnut Shop**

| Year in Business | Total Sales |
|:---:|:---:|
| 1 | $55,000 |
| 4 | $85,000 |

3.

**Jim's Depreciation on His Jet Ski**

| Years | Value |
|:---:|:---:|
| 1 | $4,500 |
| 6 | $2,500 |

2.

**Gwen's Green Beans**

| Days Growing | Height in Inches |
|:---:|:---:|
| 2 | 5 |
| 6 | 12 |

4.

**Stepping on the Brakes**

| Seconds | MPH |
|:---:|:---:|
| 2 | 51 |
| 5 | 18 |

## 9.8 Graphing Linear Data

Many types of data are related by a constant ratio. As you learned on the previous page, this type of data is linear. The slope of the line described by linear data is the ratio between the data. It is also called the **rate of change**. Plotting linear data with a constant ratio can be helpful in finding additional values.

**Example 13:** A department store prices socks per pair. Each pair of socks costs $0.75. Plot pairs of socks versus price on a Cartesian plane.

**Step 1:** Since the price of the socks is constant, you know that one pair of socks costs $0.75, 2 pairs of socks cost $1.50, 3 pairs of socks cost $2.25, and so on. Make a list of a few points.

| Pair(s) $x$ | Price $y$ |
|:---:|:---:|
| 1 | 0.75 |
| 2 | 1.50 |
| 3 | 2.25 |

**Step 2:** Plot these points on a Cartesian plane, and draw a straight line through the points.

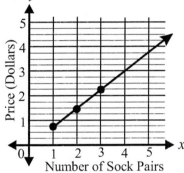

**Example 14:** What is the slope of the data in the example above? What does the slope describe?

**Solution:** You can determine the slope either by the graph or by the data points. For this data, the slope is 0.75. Remember, slope is rise/run. For every $0.75 going up the $y$-axis, you go across one pair of socks on the $x$-axis. The slope describes the price per pair of socks.

**Example 15:** Use the graph created in the above example to answer the following questions. How much would 5 pairs of socks cost? How many pairs of socks could you purchase for $3.00? Extending the line gives useful information about the price of additional pairs of socks.

**Solution 1:** The line that represents 5 pairs of socks intersects the data line at $3.75 on the $y$-axis. Therefore, 5 pairs of socks would cost $3.75.

**Solution 2:** The line representing the value of $3.00 on the $y$-axis intersects the data line at 4 on the $x$-axis. Therefore, $3.00 will buy exactly 4 pairs of socks.

**Use the information given to make a line graph for each set of data, and answer the questions related to each graph.**

1. The diameter of a circle versus the circumference of a circle is a constant ratio. Use the data given below to graph a line to fit the data. Extend the line, and use the graph to answer the next question.

**Circle**

| Diameter | Circumference |
|----------|---------------|
| 4 | 12.56 |
| 5 | 15.70 |

2. Using the graph of the data in question 1, estimate the circumference of a circle that has a diameter of 3 inches.

3. If the circumference of a circle is 3 inches, about how long is the diameter?

4. What is the slope of the line you graphed in question 1?

5. What does the slope of the line in question 4 describe?

6. The length of a side on a square and the perimeter of a square are constant ratios to each other. Use the data below to graph this relationship.

**Square**

| Length of side | Perimeter |
|----------------|-----------|
| 2 | 8 |
| 3 | 12 |

7. Using the graph from question 6, what is the perimeter of a square with a side that measure 4 inches?

8. What is the slope of the line graphed in question 6?

9. Conversions are often constant ratios. For example, converting from pounds to ounces follows a constant ratio. Use the data below to graph a line that can be used to convert pounds to ounces.

**Measurement Conversion**

| Pounds | Ounces |
|--------|--------|
| 2 | 32 |
| 4 | 64 |

10. Use the graph from question 9 to convert 40 ounces to pounds.

11. What does the slope of the line graphs for question 9 represent?

12. Graph the data below, and create a line that shows the conversion from weeks to days.

**Time**

| Weeks | Days |
|-------|------|
| 1 | 7 |
| 2 | 14 |

13. About how many days are in $2\frac{1}{2}$ weeks?

# Chapter 9 Review

1. Graph the equation $y = -\frac{1}{2}x^2 + 1$.

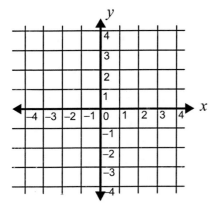

2. What is the name of the graph described by the equation $y = 2x^2 - 1$?

3. The graph of the line $y = 3x - 1$ is shown below. On the same graph, draw the line $y = -\frac{1}{3}x - 1$.

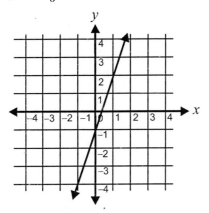

4. What is the equation of a line that is perpendicular to the line $3x + 2y = 6$ and passes through the point $(12, -15)$?

5. What is the equation of a line that is parallel to the line $-5x + y = -4$ and passes through the point $(-1, 7)$?

6. If you change the slope of the line $2x - y = 4$ to $-1$, how will the graph of the line be affected?

**Graph the equations below on a Cartesian plane.**

7. $y = -x^2 + 4$

8. $y = 3x^2 - 1$

9. $y = 4 - x^2$

10. $y = 3x^2 + 6$

**Find the $x$-intercept(s) and $y$-intercept of the following quadratic equations.**

11. $y = 2x^2 - 8$

12. $y = -x^2 - 9$

13. $y = 5x^2 + 1$

14. $y = x^2 - 6$

15. Paulo turned on the oven to preheat it. After one minute, the oven temperature was $200°$. After 2 minutes, the oven temperature was $325°$.

### Oven Temperature

| Minutes | Temperature |
|---------|-------------|
| 1 | $200°$ |
| 2 | $325°$ |

Assuming the oven temperature rose at a constant rate, write an equation that fits the data.

16. Write an equation that fits the data given below. Assume the data is linear.

### Plumber Charges per Hour

| Hour | Charge |
|------|--------|
| 1 | $170 |
| 2 | $220 |

17. The data given below show conversions between miles per hour and kilometers per hour. Based on this data, graph a conversion line on the Cartesian plane below.

**Speed**

| MPH | KPH |
|-----|-----|
| 5   | 8   |
| 10  | 16  |

18. What would be the approximate conversion of 9 mph to kph?

19. What would be the approximate conversion of 13 kph to mph?

20. A bicyclist travels 12 mph downhill. Approximately how many kph is the bicyclist traveling?

21. Use the data given below to graph the interest rate versus the interest rate on $80.00 in one year.

**$80.00 Principal**

| Interest Rate | Interest - 1 Year |
|---------------|-------------------|
| 5%            | $4.00             |
| 10%           | $8.00             |

22. About how much interest would accrue in one year at an 8% interest rate?

23. What is the slope of the line describing interest versus interest rate?

24. What information does the slope give in problem 23?

**Graph each function.**

25. $f(x) = -3\sqrt{x}$

26. $f(x) = \frac{1}{4}x^3$

27. $f(x) = -\frac{1}{8}\sqrt{x}$

# Chapter 9 Test

1. What happens to a graph if the slope changes from 2 to $-2$?

   A The graph will move down 4 spaces.
   B The graph will slant from upper left to lower right.
   C The graph will flatten out to be more vertical.
   D The graph will slant from upper right to lower left.

2. What happens to a graph if the $y$-intercept changes from 4 to $-2$?

   A The graph will move down 2 spaces.
   B The graph will slant towards the left instead of the right.
   C The graph will move down 6 spaces.
   D The graph will move up 6 spaces.

3. What are the $x$-intercept(s) of the equation $y = -2x^2 + 8$?

   A $(-2, 0)$ and $(2, 0)$
   B $(-4, 0)$ and $(4, 0)$
   C $(8, 0)$
   D There are no $x$-intercepts.

4. The graph of which pair of equations below will be parallel?

   A $x + 4y = 3$
   $3x + 4y = 3$

   B $x - 4y = 3$
   $4y - x = -3$

   C $2x - 8 = 2y$
   $2x + 8 = 2y$

   D $6x + 6 = 6y$
   $11x - 12 = 7y$

5. What are the $x$-intercept(s) of the equation $y = x^2 + 9$

   A $(9, 0)$
   B $(-3, 0)$ and $(3, 0)$
   C $(-9, 0)$ and $(9, 0)$
   D There are no $x$-intercepts.

6. Which of the following graphs represents $y = 2x^2$?

   A
   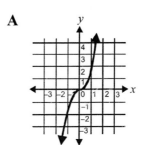

   B
   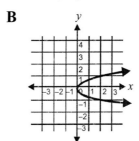

   C
   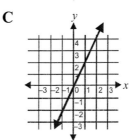

   D
   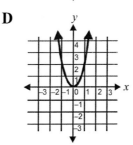

7. Which of the following statements is an accurate comparison of the lines $y = 3x - 1$ and $y = -\frac{1}{3}x - 1$?

A Only their $y$-intercepts are different.

B Only their slopes are different.

C Both their $y$-intercepts and their slopes are different.

D There is no difference between these two lines.

8. Which of the following is an equation of a line that is perpendicular to the line $l$ in the graph?

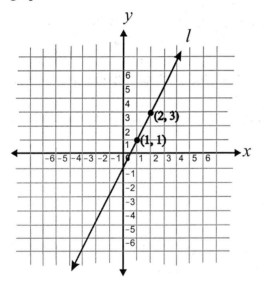

A $x - 2y = -4$

B $x - 2y = 4$

C $x + 2y = 4$

D $2x + y = 4$

9. Florence follows Great Aunt Emma's instructions for making coffee in various size coffee urns.

| Capacity of the Coffee Urn | Number of Scoops of Coffee |
|---|---|
| 4 quarts | 15 scoops |
| 6 quarts | 25 scoops |
| 10 quarts | 40 scoops |

Which of these graphs correctly plots the number of scoops as a function of the capacity of the urn?

A
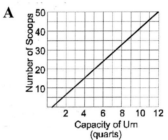

B
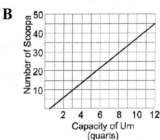

C
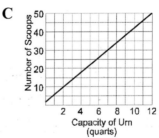

D
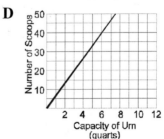

# Chapter 10
# Pairs of Linear Equations and Inequalities

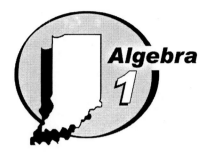

This chapter covers the following IN Algebra I standards:

| Standard 5: | Pairs of Linear Equations and Inequalities | A1.5.1 |
| | | A1.5.2 |
| | | A1.5.3 |
| | | A1.5.4 |
| | | A1.5.5 |
| | | A1.5.6 |
| Standard 9: | Mathematical Reasoning and Problem Solving | A1.9.4 |

## 10.1 Pairs of Linear Equations

Two linear equations considered at the same time are called a **pair** of linear equations. The graph of a linear equation is a straight line. The graphs of two linear equations can show that the lines are **parallel, intersecting**, or **collinear**. Two lines that are **parallel** will never intersect and have no ordered pairs in common. If two lines are **intersecting**, they have one point in common, and in this chapter, you will learn to find the ordered pair for that point. If the graph of two linear equations is the same line, the lines are said to be **collinear**.

If you are given a pair of two linear equations, and you put both equations in slope-intercept form, you can immediately tell if the graph of the lines will be **parallel, intersecting**, or **collinear**.

If two linear equations have the same slope and the same $y$-intercept, then they are both equations for the same line. They are called **collinear** or **coinciding** lines. A line is made up of an infinite number of points extending infinitely far in two directions. Therefore, collinear lines have an infinite number of points in common.

**Example 1:**    $2x + 3y = -3$    **In slope intercept form:**    $y = -\frac{2}{3}x - 1$

$\phantom{Example 1:}$    $4x + 6y = -6$    **In slope intercept form:**    $y = -\frac{2}{3}x - 1$

**The slope and $y$-intercept of both lines are the same.**

If two linear equations have the same slope but different $y$-intercepts, they are **parallel** lines. Parallel lines never touch each other, so they have no points in common.

If two linear equations have different slopes, then they are intersecting lines and share exactly one point in common.

The chart below summarizes what we know about the graphs of two equations in slope-intercept form.

| $y$-Intercepts | Slopes | Graphs | Number of Solutions |
|---|---|---|---|
| same | same | collinear | infinite |
| different | same | distinct parallel lines | none (they never touch) |
| same or different | different | intersecting lines | exactly one |

**For the pairs of equations below, put each equation in slope-intercept form, and tell whether the graphs of the lines will be collinear, parallel, or intersecting.**

1. $3y = 2x + 9$
   $18 = 6y - 4x$

2. $-x + y = -5$
   $x - y = 5$

3. $y = 3x + 2$
   $y - 3x = 2$

4. $-x = y$
   $-x = 2 + y$

5. $x + y = 4$
   $-x + y = 4$

6. $3x = y + 1$
   $y = 3x + 1$

7. $2x - y = 4$
   $-4x + 2y = -8$

8. $3x + y = 1$
   $x + y = 1$

9. $-y = x - 7$
   $y + x = -7$

10. $10x - 5y = 3$
    $5x - 10y = 3$

11. $-2x + 3y = 5$
    $x = 2 - y$

12. $4x - 3y = 12$
    $y = \frac{4}{3}x - 4$

13. $2x + 2y = 18$
    $y + x = 9$

14. $3x - 7y = 10$
    $6x - 14y = 20$

15. $2x = 4y - 1$
    $7y = x - 7$

16. $8y = x - 5$
    $y - \frac{1}{8}x = 12$

17. $3x - y = 1$
    $2y = -6x + 5$

18. $9 = 3x - y$
    $x = y + 3$

19. $-2x = y - 5$
    $x - 5 = 2y$

20. $\frac{1}{2}x = y$
    $-y = -\frac{1}{2}x$

## 10.2 Finding Common Solutions for Intersecting Lines

When two lines intersect, they share exactly one point in common.

**Example 2:** $4x + y = 3$ and $x - y = 1$

Put each equation in slope-intercept form.

$$4x + y = 3 \qquad\qquad x - y = 1$$
$$y = -4x + 3 \qquad\qquad -y = -x + 1$$
$$\qquad\qquad\qquad\qquad y = x - 1$$

slope-intercept form

Straight lines with different slopes are **intersecting lines**. Look at the graphs of the lines on the same Cartesian plane.

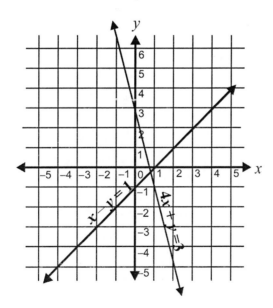

You can see from looking at the graph that the intersecting lines share one point in common. However, it is hard to tell from looking at the graph what the coordinates are for the point of intersection. To find the exact point of intersection, you can use the **substitution method** to solve the pair of equations algebraically.

## 10.3   Solving Pairs of Equations by Substitution

You can solve pairs of equations by using the substitution method.

**Example 3:**   Find the point of intersection of the following two equations:

Equation 1:   $x - y = 1$
Equation 2:   $x + 3y = 9$

**Step 1:**   Solve one of the equations for $x$ or $y$. Let's choose to solve equation 1 for $x$.

Equation 1:   $x - y = 1$
$$x = y + 1$$

**Step 2:**   Substitute the value of $x$ from equation 1 in place of $x$ in equation 2.

Equation 2:   $x + 3y = 9$
$$(y + 1) + 3y = 9$$
$$4y + 1 = 9$$
$$4y = 9 - 1$$
$$4y = 8$$
$$y = 2$$

**Step 3:**   Substitute the solution for $y$ back in equation 1 and solve for $x$.

Equation 1:   $x - y = 1$
$$x - 2 = 1$$
$$x = 3$$

**Step 4:**   The solution set is $(3, 2)$. Substitute in one or both of the equations to check.

Equation 1:   $x - y = 1$         Equation 2:   $x + 3y = 9$
$\phantom{xxxxxxx}3 - 2 = 1$ $\phantom{xxxxxxxxxxxx}3 + 3(2) = 9$
$\phantom{xxxxxxx}1 = 1$ $\phantom{xxxxxxxxxxxx}3 + 6 = 9$
$\phantom{xxxxxxxxxxxxxxxxxxxxxxxxxxxxxx}9 = 9$

The point $(3, 2)$ is common for both equations. This is the **point of intersection**.

**For each of the following pairs of equations, find the point of intersection, the common solution, using the substitution method.**

1. $x + y = 22$
   $x - y = 8$

2. $y - 5 = x$
   $y - 7 = 2x$

3. $y = 2 - x$
   $5 = y + 2x$

4. $y - 9 = -x$
   $-7 = y - 3x$

5. $y = x + 2$
   $y + 1 = 2x$

6. $5y - 8 = x$
   $3y - 4 = x$

7. $2y + 6 = x$
   $y - 3x = -13$

8. $y + 3 = 2x$
   $y + 5 = 3x$

9. $y - \frac{4}{5}x = 1$
   $y - x = 1$

10. $y + \frac{1}{3}x = -1$
    $y + 1 = 2x$

11. $y - 5 = 2x$
    $3y + x = 8$

12. $x - y = 0$
    $y + 1 = 2x$

13. $x - y = 1$
    $\frac{1}{2}x + y = 5$

14. $y + x = 1$
    $y + 4x = 10$

15. $y + x = -1$
    $y - 2x = -16$

## 10.4   Solving Pairs of Equations by Adding or Subtracting

You can solve pairs of equations algebraically by adding or subtracting an equation from another equation or pair of equations.

**Example 4:**   Find the point of intersection of the following two equations:
Equation 1: $x + y = 10$
Equation 2: $-x + 4y = 5$

**Step 1:**   Eliminate one of the variables by adding the two equations together. Since the $x$ has the same coefficient in each equation, but opposite signs, it will cancel nicely by adding.

$$
\begin{array}{ll}
x + y = 10 & \\
\underline{+ (-x + 4y = 5)} & \text{Add each like term together.} \\
0 + 5y = 15 & \text{Simplify.} \\
5y = 15 & \text{Divide both sides by 5.} \\
y = 3 &
\end{array}
$$

**Step 2:**   Substitute the solution for $y$ back into an equation, and solve for $x$.

$$
\begin{array}{lll}
\text{Equation 1:} & x + y = 10 & \text{Substitute 3 for } y. \\
& x + 3 = 10 & \text{Subtract 3 from both sides.} \\
& x = 7 &
\end{array}
$$

**Step 3:**   The solution set is $(7, 3)$. To check, substitute the solution into both of the original equations.

$$
\begin{array}{llll}
\text{Equation 1:} & x + y = 10 & \text{Equation 2:} & -x + 4y = 5 \\
& 7 + 3 = 10 & & -(7) + 4(3) = 5 \\
& 10 = 10 & & -7 + 12 = 5 \\
& & & 5 = 5
\end{array}
$$

The point $(7, 3)$ is the point of intersection.

**Example 5:**   Find the point of intersection of the following two equations:
Equation 1: $3x - 2y = -1$
Equation 2: $-4y = -x - 7$

**Step 1:**   Put the variables in equation 2 on the same side.

$$
\begin{array}{ll}
-4y = -x - 7 & \text{Add } x \text{ to both sides.} \\
x - 4y = -x + x - 7 & \text{Simplify.} \\
x - 4y = -7 &
\end{array}
$$

**Step 2:**   Add the two equations together to cancel one variable. Since each variable has the same sign and different coefficients, we have to multiply one equation by a negative number so one of the variables will cancel. Equation 1's $y$ variable has a coefficient of 2, and if multiplied by $-2$, the $y$ will have the same variable as the $y$ in equation 2, but a different sign. This will cancel nicely when added.

$$
\begin{array}{ll}
-2(3x - 2y = -1) & \text{Multiply by } -2. \\
-6x + 4y = 2 &
\end{array}
$$

**Step 3:** Add the two equations.

$$-6x + 4y = 2$$
$$\underline{+ (x - 4y = -7)}$$ Add equation 2 to equation 1.
$$-5x + 0 = -5$$ Simplify.
$$-5x = -5$$ Divide both sides by $-5$.
$$x = 1$$

**Step 4:** Substitute the solution for $x$ back into an equation and solve for $y$.

Equation 1:
$$3x - 2y = -1$$ Substitute 1 for $x$.
$$3(1) - 2y = -1$$ Simplify.
$$3 - 2y = -1$$ Subtract 3 from both sides.
$$3 - 3 - 2y = -1 - 3$$ Simplify.
$$-2y = -4$$ Divide both sides by $-2$.
$$y = 2$$

**Step 5:** The solution set is $(1, 2)$. To check, substitute the solution into both of the original equations.

Equation 1:
$$3x - 2y = -1$$
$$3(1) - 2(2) = -1$$
$$3 - 4 = -1$$
$$-1 = -1$$

Equation 2:
$$-4y = -x - 7$$
$$-4(2) = -1 - 7$$
$$-8 = -8$$

The point $(1, 2)$ is the point of intersection.

**For each of the following pairs of equations, find the point of intersection by adding the two equations together.**

1. $x + 2y = 8$
   $-x - 3y = 2$

2. $x - y = 5$
   $2x + y = 1$

3. $x - y = -1$
   $x + y = 9$

4. $3x - y = -1$
   $x + y = 13$

5. $-x + 4y = 2$
   $x + y = 8$

6. $x + 4y = 10$
   $x + 7y = 16$

7. $2x - y = 2$
   $4x - 9y = -3$

8. $x + 3y = 13$
   $5x - y = 1$

9. $-x = y - 1$
   $x = y - 1$

10. $x - y = 2$
    $2y + x = 5$

11. $5x + 2y = 1$
    $4x + 8y = 20$

12. $3x - 2y = 14$
    $x - y = 6$

13. $2x + 3y = 3$
    $3x + 5y = 5$

14. $x - 4y = 6$
    $-x - y = -1$

15. $x = 2y + 3$
    $y = 3 - x$

## 10.5  Graphing Pairs of Inequalities

Pairs of inequalities are best solved graphically. Look at the following example.

**Example 6:**  Sketch the solution set of the following pair of inequalities:

$$y > -2x - 1 \text{ and } y \le 3x$$

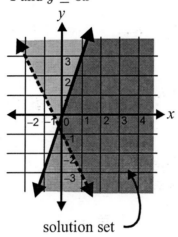

solution set

**Step 1:**  Graph both inequalities on a Cartesian plane. Study the chapter on graphing inequalities if you need to review.

**Step 2:**  Shade the portion of the graph that represents the solution set to each inequality just as you did in the chapter on graphing inequalities.

**Step 3:**  Any shaded region that overlaps is the solution set of both inequalities.

**Graph the following pairs of inequalities on your own graph paper. Shade and identify the solution set for both inequalities.**

1.  $9y \le 6x + 18$
    $-4x - 4y \ge 8$

2.  $14x + 14y > 42$
    $8x \ge 24 + 12y$

3.  $x + y > -3$
    $-6x + 12y < -12$

4.  $x - y \ge 2$
    $2y \le -4x + 4$

5.  $4x > 3y - 12$
    $-\frac{2}{3}x \ge y - 2$

6.  $2y < -6x + 4$
    $y - 2x < -2$

7.  $y < \frac{1}{3}x - 1$
    $-3x < y - 3$

8.  $-2x + y > 2$
    $y \le -x - 3$

9.  $y + \frac{2}{3}x \ge 3$
    $y - 2 < 2x$

10.  $y + \frac{3}{2}x \ge -3$
    $x + y \ge -2$

11.  $y \le x - 2$
    $y \le -x - 3$

12.  $y - 2x \le -1$
    $y + 2 > -x$

## 10.6    Solving Word Problems with Pairs of Equations

Certain word problems can be solved using systems of equations.

**Example 7:**    In a game show, Andre earns 6 points for every right answer and loses 12 points for every wrong answer. He has answered correctly 12 times as many as he has missed. His final score was 120. How many times did he answer correctly?

**Step 1:**    Let $r$ = number of right answers. Let $w$ = number of wrong answers.

We know 2 sets of information that can be made into equations with 2 variables.

He earns +6 points for right answers and loses 12 points for wrong answers.

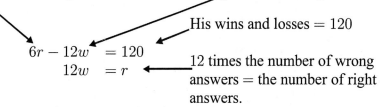

His wins and losses = 120

$$6r - 12w = 120$$
$$12w = r$$

12 times the number of wrong answers = the number of right answers.

**Step 2:**    Substitute the value for $r$ ($12w$) in the first equation.

$$6(12w) - 12w = 120$$
$$w = 2$$

**Step 3:**    Substitute the value for $w$ back in the equation.

$$6r - 12(2) = 120$$
$$r = 24$$

**Example 8:**    Ms. Sudberry bought pencils and stickers for her first grade class on two different days. The pencils and stickers cost the same each time she went to the store. How much did she pay for each pencil?

|         | Pencils | Stickers | Total Cost |
|---------|---------|----------|------------|
| Tuesday | 30      | 40       | $47.50     |
| Saturday| 60      | 5        | $20.00     |

**Step 1:**    Set up your two equations. Let the price of pencils equal $x$, and the price of stickers equal $y$.

The amount of the pencils times the price of pencils ($x$) plus the amount of the stickers times the price of stickers ($y$) equals the total amount paid for both pencils and stickers.

Equation 1: $30x + 40y = \$47.50$
Equation 2: $60x + 5y = \$20.00$

**Step 2:**    Solve the equations by using one of the methods taught in this chapter. We will use the adding and subtracting method. First, multiply equation 1 by $-2$, so $x$ will have the same coefficient in each equation but with opposite signs.

$$-2(30x + 40y = \$47.50) = -60x - 80y = -\$95.00$$

**Step 3:**    Add the new equation 1 to equation 2.

$$
\begin{array}{rrrrl}
-60x & - & 80y & = & -\$95.00 \\
+\quad 60x & + & 5y & = & \$20.00 \\
\hline
0x & - & 75y & = & -\$75.00
\end{array}
$$

The new equation is $-75y = -\$75.00$.

**Step 4:**    Solve for $y$.
$-75y = -\$75.00$
$y = \$1.00$
Now, we know the price of stickers, but the question asked for the price of each pencil.

**Step 5:**    Substitute the value of $y$ into either equation and solve for $x$ to find the price of each pencil.
$30x + 40y = \$47.50$
$30x + 40\,(\$1.00) = \$47.50$
$30x + \$40.00 = \$47.50$
$30x = \$7.50$
$x = \$0.25$
The cost of each pencil is $\$0.25$.

**Use systems of equations to solve the following word problems.**

1. The sum of two numbers is 140 and their difference is 20. What are the two numbers?

2. The sum of two numbers is 126 and their difference is 42. What are the two numbers?

3. Kayla gets paid $6.00 for raking leaves and $8.00 for mowing the lawn of each of the neighbors in her subdivision. This year she mowed the lawns 12 times more than she raked leaves. In total, she made $918.00 for doing both. How many times did she rake the leaves?

4. Prices for the movie are $4.00 for children and $8.00 for adults. The total amount of ticket sales is $1,176. There are 172 tickets sold. How many adults and children buy tickets?

5. A farmer sells a dozen eggs at the market for $2.00 and one of his bags of grain for $5.00. He has sold 5 times as many bags of grain as he has dozens of eggs. By the end of the day, he has made $243.00 worth of sales. How many bags of grain did he sell?

6. Every time Lauren does one of her chores, she gets 15 minutes to talk on the phone. When she does not perform one of her chores, she gets 20 minutes of phone time taken away. This week she has done her chores 5 times more than she has not performed her chores. In total, she has accumulated 165 minutes. How many times has Lauren not performed her chores?

7. The choir sold boxes of candy and teddy bears near Valentine's Day to raise money. They sold twice as many boxes of candy as they did teddy bears. Bears sold for $8.00 each and candy sold for $6.00. They collected $380. How much of each item did they sell?

8. Mr. Marlow keeps ten and twenty dollar bills in his dresser drawer. He has 1 less than twice as many tens as twenties. He has $550 altogether. How many ten dollar bills does he have?

9. Kosta is a contestant on a math quiz show. For every correct answer, Kosta receives $18.00. For every incorrect answer, Kosta loses $24.00. Kosta answers the questions correctly twice as often as he answers the questions incorrectly. In total, Kosta wins $72.00. How many questions does Kosta answer incorrectly?

10. John Vasilovik works in landscaping. He gets paid $50 for each house he pressure-washes and $20 for each lawn he mows. He gets 4 times more jobs for mowing lawns than for pressure-washing houses. During a given month, John earns $2,600. How many houses does John pressure wash?

11. Every time Stephen walks the dog, he gets 30 minutes to play video or computer games. When he does not take out the dog on time, he gets a mess to clean up and loses 1 hour of video or computer game time. This week he has walked the dog on time 8 times more than he did not walk the dog on time. In total, he has accumulated 3 hours of video or computer time. How many times has Stephen not walked the dog on time?

12. On Friday, Rosa bought party hats and kazoos for her friend's birthday party. On Saturday she decided to purchase more when she found out more people were coming. How much did she pay for each party hat?

|  | Hats | Kazoos | Total Cost |
|---|---|---|---|
| Friday | 15 | 20 | $15.00 |
| Saturday | 10 | 5 | $8.75 |

13. Timothy and Jesse went to purchase sports clothing they needed to play soccer. The table below shows what they bought and the amount they paid. What is the price of 1 soccer jersey?

|  | Soccer Jerseys | Tube Socks | Total Cost |
|---|---|---|---|
| Timothy | 4 | 7 | $78.30 |
| Jesse | 3 | 5 | $57.60 |

# Chapter 10 Review

**For each pair of equations below, tell whether the graphs of the lines are collinear, parallel, or intersecting.**

1. $y = \frac{1}{2}x + 4$
   $y = \frac{1}{2}x - 3$

2. $3 - x = y$
   $2x + y = 1$

3. $\frac{1}{2} = 2x - \frac{1}{2}y$
   $4x = y + 1$

4. $y = 5x + 5$
   $y = 1 + 5x$

5. $y - \frac{1}{8}x = 2$
   $8y = 16 + x$

6. $3x - y = 2$
   $x + y = 2$

**Find the common solution for each of the following pairs of equations, using the substitution method.**

7. $2x - y = 3$
   $3x + 4y = -1$

8. $x + 2y = 1$
   $x + 3y = 1$

9. $-y = -2x + 7$
   $-x = -2y - 2$

10. $3x + 6y = 18$
    $4y - 2x = 12$

11. $2x = y - 10$
    $-\frac{1}{2}x = y$

12. $-2x + y = -6$
    $2x - 2y = 16$

**Graph the following pairs of inequalities on your own graph paper. Shade and identify the solution set for both inequalities.**

13. $y - 2x \geq -4$
    $y + 2x \leq 1$

14. $y < 3x + 2$
    $y - \frac{1}{2}x > -1$

15. $4y \leq -3x - 12$
    $2y \leq x + 6$

16. $y - 2x \leq 4$
    $4y + 3x \leq -12$

17. $y \leq \frac{1}{2}x + 1$
    $y \leq 3x + 3$

18. $y \leq \frac{3}{2}x - 3$
    $y \geq 2x - 4$

**Find the point of intersection for each pair of equations by adding and/or subtracting the two equations.**

19. $2x + y = 4$
    $3x - y = 6$

20. $x + 2y = 3$
    $x + 5y = 0$

21. $x + y = 1$
    $y = x + 7$

22. $2x + 4y = 5$
    $3x + 8y = 9$

23. $2x - 2y = 7$
    $3x - 5y = \frac{5}{2}$

24. $x - 3y = -2$
    $y = -\frac{1}{3}x + 4$

**Use pairs of equations to solve the following word problems.**

25. Hargrove High School sold 227 tickets for their last basketball game. Adult tickets sold for $5 and student tickets were $2. How many adult tickets were sold if the ticket sales totalled $574?

26. Zack is an ostrich and llama breeder. He sells full-grown ostriches for $625 and full-grown llamas for $750 each. Zack sold 1 less than 3 times as many llamas as ostriches this year. His total sales for the year were $7,875.00. How many llamas did Zack sell during this year?

27. Sarah and Abdul played Geography Quiz Bowl during summer school. For every time Abdul got an answer right, Sarah got 4 answers right. If Sarah and Abdul correctly answered 75 questions, how many times did Abdul answer correctly?

# Chapter 10 Test

1. Consider the following equations:

    $f(x) = 6x + 2$ and $f(x) = 3x + 2$

    Which of the following statements is true concerning the graphs of these equations?

    **A**  The lines are collinear.

    **B**  The lines intersect at exactly one point.

    **C**  The lines are parallel to each other.

    **D**  The graphs of the lines intersect each other at the point $(2, 2)$.

2. What is the solution to the following pair of equations?

    $y = 4x - 8$
    $y = 2x$

    **A**  $(-4, -8)$
    **B**  $(4, 8)$
    **C**  $(-1, -2)$
    **D**  $(1, 2)$

3. Two lines are shown on the grid. One line passes through the origin and the other passes through $(-1, -1)$ with a $y$-intercept of 2. Which pair of equations below the grid identifies these lines?

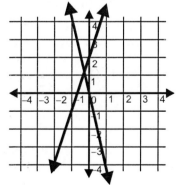

    **A**  $y = \frac{1}{4}x$ and $y = \frac{1}{3}x + 2$
    **B**  $x - 2y = 6$ and $4x + y = 4$
    **C**  $y = 4x$ and $y = \frac{1}{3}x$
    **D**  $y = 3x + 2$ and $y = -4x$

4. The graph of which pair of equations below is parallel?

    **A**  $x + 3y = 3$
    $\quad\ 3x + y = 3$
    **B**  $3x + 3y = 6$
    $\quad\ 9x - 3y = 6$
    **C**  $x - 3y = 6$
    $\quad\ 3y - x = -3$
    **D**  $x + 3 = y$
    $\quad\ x - 3 = 2y$

5. Which ordered pair is a solution for the following pair of equations?

    $-3x + 7y = 25$
    $3x + 3y = -15$

    **A**  $(-13, -2)$
    **B**  $(-6, 1)$
    **C**  $(-3, -2)$
    **D**  $(-20, -5)$

6. For the following pair of equations, find the point of intersection (common solution) using the substitution method.

    $-3x - y = -2$
    $5x + 2y = 20$

    **A**  $(2, -4)$
    **B**  $(2, 5)$
    **C**  $(-16, 50)$
    **D**  $\left(\frac{1}{5}, \frac{1}{2}\right)$

7. What is the intersection point of the graphs of the equations $2x - y = 2$ and $4x - 9y = -3$?

    **A.**  $\left(\frac{1}{2}, -1\right)$
    **B.**  $\left(1, \frac{3}{2}\right)$
    **C.**  $\left(-\frac{3}{2}, 1\right)$
    **D.**  $\left(\frac{3}{2}, 1\right)$

8. What is the intersection point of the graphs of the equations $x - 4y = 6$ and $-x - y = -1$?

   A. $(2, 1)$
   B. $(2, -1)$
   C. $(-2, 1)$
   D. $(-2, -1)$

9. The graphs of the equations $x = -y$ and $x = 4 - y$ are

   A  collinear.
   B  parallel.
   C  intersecting.
   D  not enough information.

10. The graphs of the equations $x + y = 2$ and $5x + 5y = 10$ are

   A  collinear.
   B  parallel.
   C  intersecting.
   D  not enough information.

11. What is the intersection point the graphs of the equations $x = y + 3$ and $y = 3 - x$?

   A  $(3, 0)$
   B  $(0, 3)$
   C  $(3, 3)$
   D  $(-3, 0)$

12. What is the intersection point of the graphs of the equations $2x + 3y = 2$ and $4x - 9y = -1$?

   A  $(2, 3)$
   B  $\left(\dfrac{1}{3}, \dfrac{1}{2}\right)$
   C  $\left(\dfrac{1}{2}, \dfrac{1}{3}\right)$
   D  $(3, 2)$

13. What is the intersection point of the graphs of the equations $-x = y - 1$ and $x = y - 1$?

   A  $(0, 1)$
   B  $(1, 0)$
   C  $(-2, -1)$
   D  $(2, 1)$

14. The admission fee at the fair is $1.50 for children and $4 for adults. On a certain day, 2,200 people enter the fair and $5,050 is collected. How many children attended? How many adults attended?

   A  children $= 750$, adults $= 1000$
   B  children $= 1000$, adults $= 700$
   C  children $= 1300$, adults $= 750$
   D  children $= 1500$, adults $= 700$

15. Which inequalities are represented by the graph?

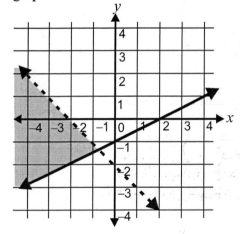

   A. $-x \geq -2y - 2$
      $-2x - 2y > 4$

   B. $2x - 2y \leq 4$
      $3x + 3y \leq -9$

   C. $-3x \leq 6 + 2y$
      $y \geq -x - 2$

   D. $3x + 4y \geq 12$
      $y > -3x + 2$

16. Which graph represents $2x + 2y \geq -4$ and $3y < 2x + 6$?

17. Which graph represents $2x - 2y \leq 4$ and $3x + 3y \leq -9$?

**A**

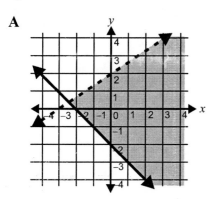

**A**

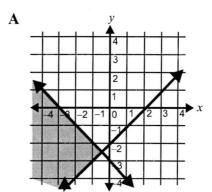

**B**

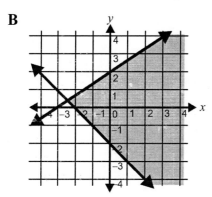

**B**

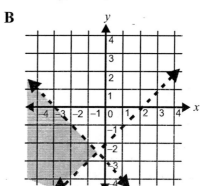

**C**

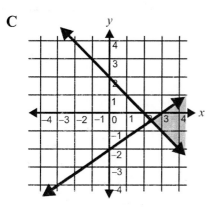

**C**

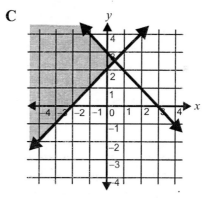

**D**

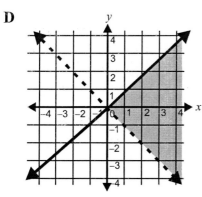

**D**

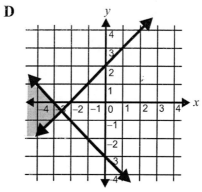

# Chapter 11
# Relations and Functions

This chapter covers the following IN Algebra I standards:

| Standard 3: | Relations and Functions | A1.3.1 |
|---|---|---|
| | | A1.3.2 |
| | | A1.3.3 |
| | | A1.3.4 |

## 11.1 Relations

A **relation** is a set of ordered pairs. We call the set of the first members of each ordered pair the **domain** of the relation. We call the set of the second members of each ordered pairs the **range**.

**Example 1:** State the domain and range of the following relation:
$$\{(2,4), (3,7), (4,9), (6,11)\}$$

**Solution:** Domain: $\{2,3,4,6\}$  the first member of each ordered pair
Range: $\{4,7,9,11\}$  the second member of each ordered pair

**State the domain and range for each relation.**

1. $\{(2,5), (9,12), (3,8), (6,7)\}$

2. $\{(12,4), (3,4), (7,12), (26,19)\}$

3. $\{(4,3), (7,14), (16,34), (5,11)\}$

4. $\{(2,45), (33,43), (98,9), (43,61), (67,54)\}$

5. $\{(78,14), (29,67), (84,49), (16,18), (98,46)\}$

6. $\{(-8,16), (23,-7), (-4,-9), (16,-8), (-3,6)\}$

7. $\{(-7,-4), (-3,16), (-4,17), (-6,-8), (-8,12)\}$

8. $\{(-1,-2), (3,6), (-7,14), (-2,8), (-6,2)\}$

9. $\{(0,9), (-8,5), (3,12), (-8,-3), (7,18)\}$

10. $\{(58,14), (44,97), (74,32), (6,18), (63,44)\}$

11. $\{(-7,0), (-8,10), (-3,11), (-7,-32), (-2,57)\}$

12. $\{(18,34), (22,64), (94,36), (11,18), (91,45)\}$

When given an equation in two variables, the domain is the set of $x$ values that satisfies the equation. The range is the set of $y$ values that satisfies the equation.

**Example 2:**   Find the range of the relation $3x = y + 2$ for the domain $\{-1, 0, 1, 2, 3\}$.
Solve the equation for each value of $x$ given. The result, the $y$ values, will be the range.

| Given: | | | Solution: | |
|---|---|---|---|---|
| $x$ | $y$ | | $x$ | $y$ |
| $-1$ | | | $-1$ | $-5$ |
| $0$ | | | $0$ | $-2$ |
| $1$ | | | $1$ | $1$ |
| $2$ | | | $2$ | $4$ |
| $3$ | | | $3$ | $7$ |

The range is $\{-5, -2, 1, 4, 7\}$.

**Find the range of each relation for the given domain.**

| | **Relation** | **Domain** | **Range** |
|---|---|---|---|
| 1. | $y = 5x$ | $\{1, 2, 3, 4\}$ | |
| 2. | $y = |x|$ | $\{-3, -2, -1, 0, 1\}$ | |
| 3. | $y = 3x + 2$ | $\{0, 1, 3, 4\}$ | |
| 4. | $y = -|x|$ | $\{-2, -1, 0, 1, 2\}$ | |
| 5. | $y = -2x + 1$ | $\{0, 1, 3, 4\}$ | |
| 6. | $y = 10x - 2$ | $\{-2, -1, 0, 1, 2\}$ | |
| 7. | $y = 3|x| + 1$ | $\{-2, -1, 0, 1, 2\}$ | |
| 8. | $y - x = 0$ | $\{1, 2, 3, 4\}$ | |
| 9. | $y - 2x = 0$ | $\{1, 2, 3, 4\}$ | |
| 10. | $y = 3x - 1$ | $\{0, 1, 3, 4\}$ | |
| 11. | $y = 4x + 2$ | $\{0, 1, 3, 4\}$ | |
| 12. | $y = 2|x| - 1$ | $\{-2, -1, 0, 1, 2\}$ | |

## 11.2   Determining Domain and Range From Graphs

The domain is all of the $x$ values that lie on the function in the graph from the lowest $x$ value to the highest $x$ value. The range is all of the $y$ values that lie on the function in the graph from the lowest $y$ to the highest $y$.

**Example 3:**     Find the domain and range of the graph.

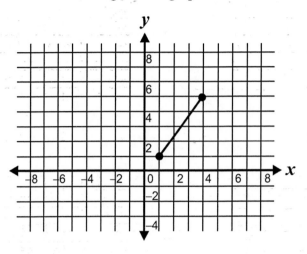

**Step 1:**     First find the lowest $x$ value depicted on the graph. In this case it is 1. Then find the highest $x$ value depicted on the graph. The highest value of $x$ on the graph is 4. The domain must contain all of the values between the lowest $x$ value and the highest $x$ value. The easiest way to write this is $1 \leq \text{Domain} \leq 4$ or $1 \leq x \leq 4$.

**Step 2:**     Perform the same process for the range, but this time look at the lowest and highest $y$ values. The answer is $1 \leq \text{Range} \leq 5$ or $1 \leq y \leq 5$.

**Find the domain and range of each graph below. Write your answers in the line provided.**

1.

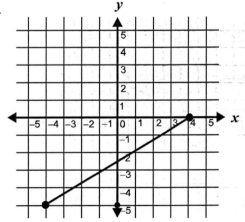

2.

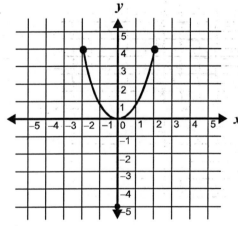

_____

_____

3.

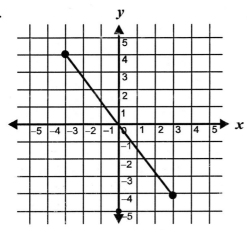

_____

4.

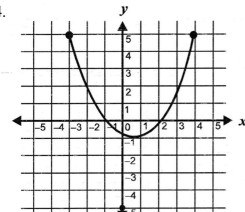

_____

5.

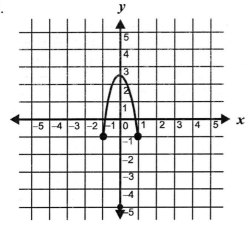

_____

6.

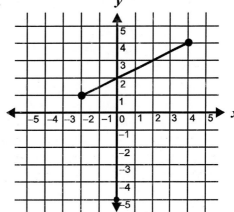

_____

7.

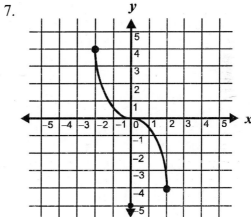

_____

8.

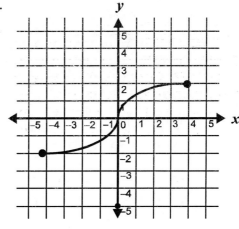

_____

## 11.3 Functions

**Some relations** are also **functions**. A relation is a function if **for every element in the domain, there is exactly one element in the range**. In other words, for each value for $x$ there is only one unique value for $y$.

**Example 4:** $\{(2,4), (2,5), (3,4)\}$ is **NOT** a function because in the first pair, 2 is paired with 4, and in the second pair, 2 is paired with 5. The 2 can be paired with only one number to be a function. In this example, the $x$ value of 2 has more than one value for $y$: 4 and 5.

**Example 5:** $\{(1,2), (3,2), (5,6)\}$ **IS** a function. Each first number is paired with only one second number. The 2 is repeated as a second number, but the relation remains a function.

**Determine whether the ordered pairs of numbers below represent a function. Write "F" if it is a function. Write "NF" if it is not a function.**

1. $\{(-1,1), (-3,3), (0,0), (2,2)\}$ _____

2. $\{(-4,-3), (-2,-3), (-1,-3), (2,-3)\}$ _____

3. $\{(5,-1), (2,0), (2,2), (5,3)\}$ _____

4. $\{(-3,3), (0,2), (1,1), (2,0)\}$ _____

5. $\{(-2,-5), (-2,-1), (-2,1), (-2,3)\}$ _____

6. $\{(0,2), (1,1), (2,2), (4,3)\}$ _____

7. $\{(4,2), (3,3), (2,2), (0,3)\}$ _____

8. $\{(-1,-1), (-2,-2), (3,-1), (3,2)\}$ _____

9. $\{(2,-2), (0,-2), (-2,0), (1,-3)\}$ _____

10. $\{(2,1), (3,2), (4,3), (5,-1)\}$ _____

11. $\{(-1,0), (2,1), (2,4), (-2,2)\}$ _____

12. $\{(1,4), (2,3), (0,2), (0,4)\}$ _____

13. $\{(0,0), (1,0), (2,0), (3,0)\}$ _____

14. $\{(-5,-1), (-3,-2), (-4,-9), (-7,-3)\}$ _____

15. $\{(8,-3), (-4,4), (8,0), (6,2)\}$ _____

16. $\{(7,-1), (4,3), (8,2), (2,8)\}$ _____

17. $\{(4,-3), (2,0), (5,3), (4,1)\}$ _____

18. $\{(2,-6), (7,3), (-3,4), (2,-3)\}$ _____

19. $\{(1,1), (3,-2), (4,16), (1,-5)\}$ _____

20. $\{(5,7), (3,8), (5,3), (6,9)\}$ _____

## 11.4   Recognizing Functions

Recall that a relation is a function with only one $y$ value for every $x$ value. We can depict functions in many ways including through graphs.

**Example 6:**

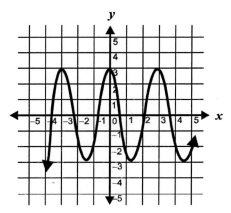

This graph **IS** a function because it has only one $y$ value for each value of $x$.

**Example 7:**

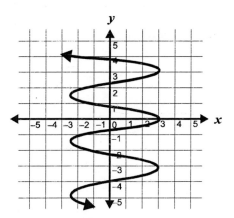

This graph is **NOT** a function because there is more than one $y$ value for each value of $x$.

**HINT:** An easy way to determine a function from a graph is to do a vertical line test. First, draw a vertical line that crosses over the whole graph. If the line crosses the graph more than one time, then it is not a function. If it only crosses it once, it is a function. Take Example 2 above:

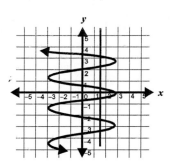

Since the vertical line passes over the graph six times, it is not a function.

**Determine whether or not each of the following graphs is a function.  If it is, write function on the line provided.  If it is not a function, write NOT a function on the line provided.**

1.

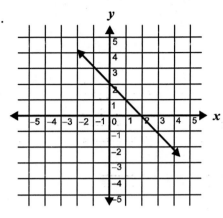

_____

4.

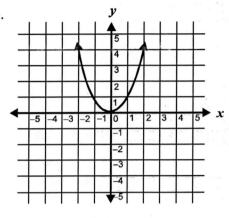

_____

2.

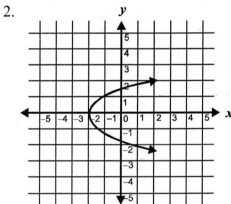

_____

5.

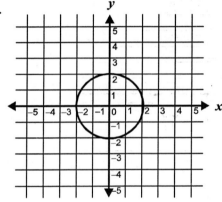

_____

3.

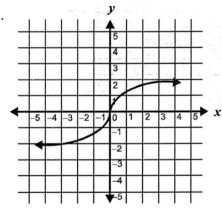

_____

6.

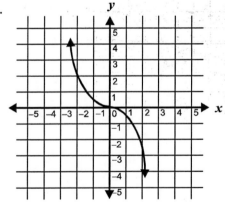

_____

7.

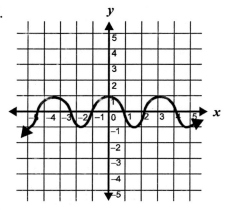

_____

8.

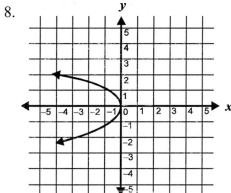

_____

9.

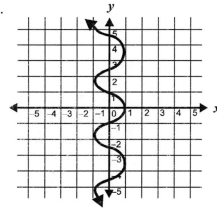

_____

10.

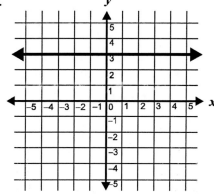

_____

11.

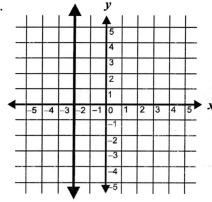

_____

12.

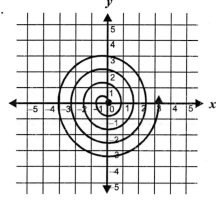

_____

## 11.5  Interpreting a Graph Representing a Given Situation

Real-world situations are sometimes modeled by graphs. Although an equation cannot be written for most of these graphs, interpreting these graphs provides valuable information. Situations may be represented on a graph as a function of time, length, temperature, etc.

The graph below depicts the temperature of a pond at different times of the day. Refer to the graph as you read through examples 1 and 2.

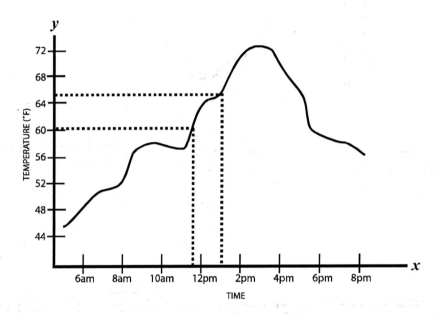

**Example 8:**  If it is known that a specific breed of fish is most active in waters between 60°F and 65°F, what time of the day would this fish be the most active in this particular pond?

To find the answer, draw lines from the 60°F and 65°F points on the $y$-axis to the graph. Then, draw vertical lines from the graph to the $x$-axis. The time range between the two vertical lines on the $x$-axis indicates the time that the fish are most active. It can be determined from the graph that the fish are most active between 11:30 am and 1:00 PM.

**Example 9:**  Describe the way the temperature of the pond acts as a function of time.

At 6:00 AM, the temperature of the pond is about 47°F. The temperature increases relatively steadily throughout the morning and early afternoon. The temperature peaks at 72°F, which is around 2:30 PM during the day. Afterwards, the temperature of the pond starts to decrease. The later it gets in the evening, the more the temperature of the water decreases. The graph shows that at 8 PM the temperature of the pond is about 57°F.

**Use the graphs to answer the questions. Circle your answers.**

The following graph depicts the number of articles of clothing as a function of time throughout the year. Use this graph for questions 1 and 2.

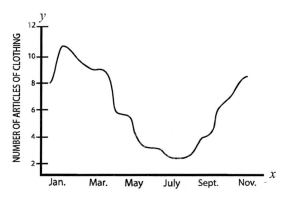

1. According to the graph, in what month are the most articles of clothing worn?

   (A) February
   (B) March
   (C) May
   (D) November

2. What is the average number of clothing a person wears in June?

   (A) 6
   (B) 3
   (C) 2
   (D) 5

The graph below depicts the efficiency of energy transfer as a function of distance in a certain element. Use the graph to answer question 3 and 4.

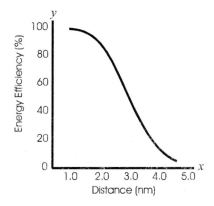

3. At what distance is the energy efficiency at 50%?

   (A) 1.0 nm
   (B) 2.0 nm
   (C) 3.5 nm
   (D) 3.0 nm

4. What is the energy efficiency at distance 2.5 nm?

   (A) 100%
   (B) 90%
   (C) 85%
   (D) 75%

**Find the best graph to match each scenario.**

1. Cathy begins her two-hour drive to her mother's house in her new sedan. She drives slowly through her city for thirty minutes to reach Interstate 95. After she gets on the highway, she travels a constant 60–70 miles per hour for the next hour until she reaches her mother's exit. She then drives slowly down back roads to arrive at her mother's house.

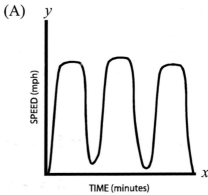

(A)

2. Phillip is flying to Texas for a business meeting. When his flight leaves, the airplane increases its speed a great deal until it reaches about 550 miles per hour. After 20 minutes, the plane levels off for the last 45 minutes at 500 miles per hour. As the airplane nears the airport in Fort Worth, TX, it decreases its speed until it lands and reaches zero miles per hour.

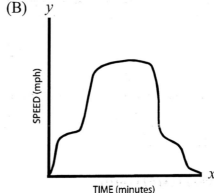

(B)

3. Erica and her father like to build rockets for fun, and every Saturday they go to the park by their house to launch the rockets. Almost immediately after takeoff, the rocket reaches its greatest speed. Affected by gravity, it slows down until it reaches its peak height. It again speeds up as it descends to the ground.

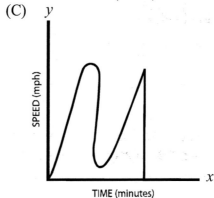

(C)

4. Molly and her mother ride the train each time they go to the zoo. Molly knows that the train slows down twice so that the passengers can view the animals. Her favorite part of the ride, though, is when the train moves very quickly before it slows down to approach the station and come to a stop.

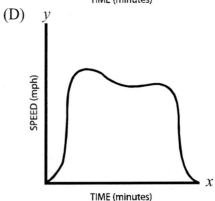

(D)

# Chapter 11 Review

1. What is the domain of the following relation?

$\{(-1, 2), (2, 5), (4, 9), (6, 11)\}$

2. What is the range of the following relation?

$\{(0, -2), (-1, -4), (-2, 6), (-3, -8)\}$

3. Find the range of the relation $y = 5x$ for the domain $\{0, 1, 2, 3, 4\}$.

4. Find the range of the relation $y = \dfrac{3(x - 2)}{5}$ for the domain $\{-8, -3, 7, 12, 17\}$.

5. Find the range of the relation $y = 10 - 2x$ for the domain $\{-8, -4, 0, 4, 8\}$.

**For each of the following relations given in questions 6–10, write F if it is a function and NF if it is not a function.**

6. $\{(1, 2), (2, 2), (3, 2)\}$

7. $\{(-1, 0), (0, 1), (1, 2), (2, 3)\}$

8. $\{(2, 1), (2, 2), (2, 3)\}$

9. $\{(1, 7), (2, 5), (3, 6), (2, 4)\}$

10. $\{(0, -1), (-1, -2), (-2, -3), (-3, -4)\}$

11. Draw the graph of the following situation on the Cartesian plane provided. A girl rode her bicycle up a hill, then coasted down the other side of the hill on her bike. At the bottom she stopped.

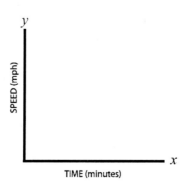

**Find the domain and range of each graph in questions 12 and 13.**

12.

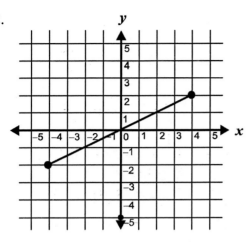

13.

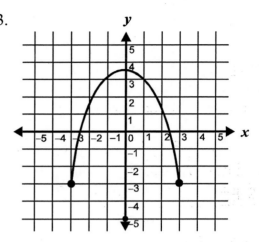

**For questions 14 and 15, determine whether or not the graphs are functions.**

14.

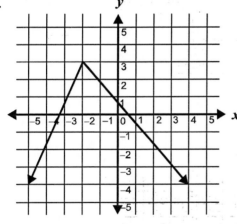

15.

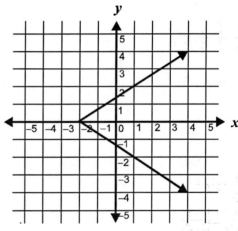

# Chapter 11 Test

1. Which of the following relations is a function?

   **A** $\{(0,-1)(1,-1)(0,-2)\}$
   **B** $\{(-1,1)(-1,-1)(0,0)\}$
   **C** $\{(2,1)(1,0)(0,-1)\}$
   **D** $\{(-1,1)(-1,0)(-1,-1)\}$

2. Find the range of the following function for the domain $\{-2,-1,0,3\}$.

$$y = \frac{2+x}{4}$$

   **A** $\left\{0, \dfrac{3}{4}, 1, \dfrac{5}{4}\right\}$

   **B** $\left\{0, \dfrac{1}{4}, \dfrac{1}{2}, \dfrac{5}{4}\right\}$

   **C** $\left\{1, -\dfrac{1}{4}, \dfrac{1}{2}, \dfrac{5}{4}\right\}$

   **D** $\left\{\dfrac{1}{4}, \dfrac{3}{4}, \dfrac{1}{2}, \dfrac{5}{4}\right\}$

3. Which of the following graphs is a function?

   **A**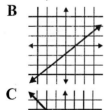

   **B**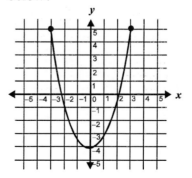

   **C**

   **D**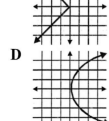

4. The following graph depicts the height of a projectile as a function of time.

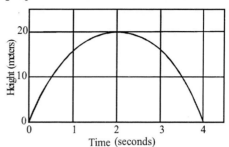

What is the domain (D) of this function?

   **A** 0 meters $\le$ D $\le$ 20 meters
   **B** 4 seconds $\le$ D $\le$ 20 meters
   **C** 20 meters $\le$ D $\le$ 4 seconds
   **D** 0 seconds $\le$ D $\le$ 4 seconds

5. What is the range of the following relation?
   $\{(1,2)(4,9)(7,8)(10,13)\}$

   **A** $\{1,4,7,10\}$
   **B** $\{2,9,8,13\}$
   **C** $\{3,13,15,23\}$
   **D** $\{1,3,1,3\}$

6. What is the domain of the function graphed below?

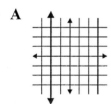

   **A** $-3$
   **B** $3$
   **C** $-4 \le x \le 5$
   **D** $-3 \le x \le 3$

7. Elisha set out walking to the bus stop. Suddenly, she realized she had forgotten her lunch box. She ran back home, found her lunch box, and ran to the bus stop so as not to miss her bus. Which of the following graphs best models this situation?

**A**

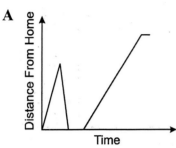

**B**

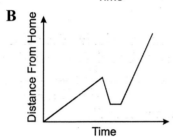

**C**

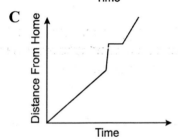

**D**

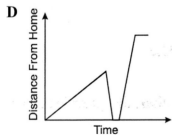

8. Which ordered pair could be part of this function: $(2, 4), (3, 7), (9, 1)$?

**A** $(4, 1)$
**B** $(3, 11)$
**C** $(2, 9)$
**D** $(9, 0)$

9. What is the range of the function graphed below?

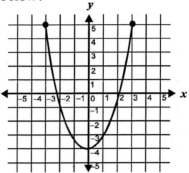

**A** $-3$
**B** $3$
**C** $-4 \leq x \leq 5$
**D** None of the above

10. Which is not a function?

**A**

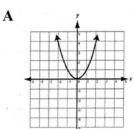

**B**

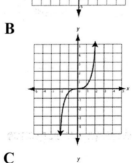

**C**

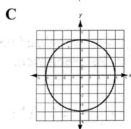

**D**

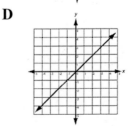

# Chapter 12
# Mathematical Reasoning

This chapter covers the following IN Algebra I standards:

| Standard 9: | Mathematical Reasoning and Problem Solving | A1.9.1 |
| --- | --- | --- |
| | | A1.9.2 |
| | | A1.9.5 |
| | | A1.9.6 |
| | | A1.9.7 |
| | | A1.9.8 |

## 12.1    Mathematical Reasoning/Logic

The IN Algebra 1 test calls for skill development in mathematical **reasoning** or **logic**. The ability to use logic is an important skill for solving math problems, but it can also be helpful in real-life situations. For example, if you need to get to Park Street, and the Park Street bus always comes to the bus stop at 3 PM, then you know that you need to get to the bus stop by at least 3 PM. This is a real-life example of using logic, which many people would call "common sense."

There are many different types of statements which are commonly used to describe mathematical principles. However, using the rules of logic, the truth of any mathematical statement must be evaluated. Below is a list of tools used in logic to evaluate mathematical statements.

**Logic** is the discipline that studies valid reasoning. There are many forms of valid arguments, but we will just review a few here.

A **proposition** is usually a declarative sentence which may be true or false.

An **argument** is a set of two or more related propositions, called **premises**, that provide support for another proposition, called the **conclusion**.

**Deductive reasoning** is an argument which begins with general premises and proceeds to a more specific conclusion. Most elementary mathematical problems use deductive reasoning.

**Inductive reasoning** is an argument in which the truth of its premises make it likely or probable that its conclusion is true.

## ARGUMENTS

Most of logic deals with the evaluation of the validity of arguments. An argument is a group of statements that includes a conclusion and at least one premise. A premise is a statement that you know is true or at least you assume to be true. Then, you draw a conclusion based on what you know or believe is true in the premise(s). Consider the following example:

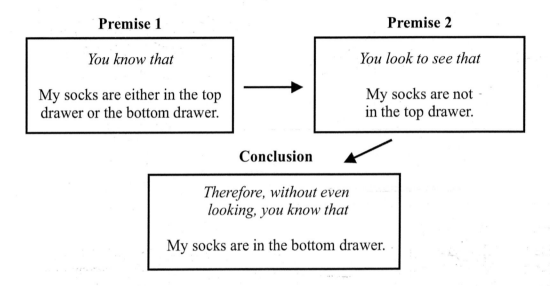

This argument is an example of deductive reasoning, where the conclusion is "deduced" from the premises and nothing else. In other words, if Premise 1 and Premise 2 are true, you don't even need to look in the bottom drawer to know that the conclusion is true.

**For numbers 1–5, what conclusion can be drawn from each proposition?**

1. All squirrels are rodents. All rodents are mammals. Therefore,

2. All fractions are rational numbers. All rational numbers are real numbers. Therefore,

3. All squares are rectangles. All rectangles are parallelograms. All parallelograms are quadrilaterals. Therefore,

4. All Chevrolets are made by General Motors. All Luminas are Chevrolets. Therefore,

5. If a number is even and divisible by three, then it is divisible by six. Eighteen is divisible by six. Therefore,

## 12.2 Deductive and Inductive Arguments

In general, there are two types of logical arguments: **deductive** and **inductive**. Deductive arguments tend to move from general statements or theories to more specific conclusions. Inductive arguments tend to move from specific observations to general theories.

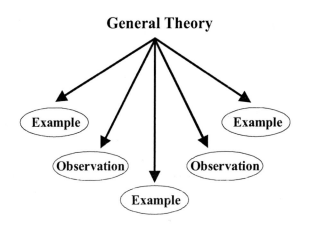

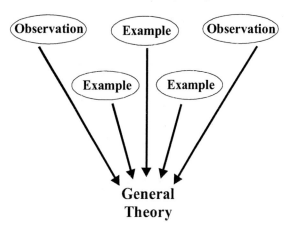

Compare the two examples below:

| **Deductive Argument** | | **Inductive Argument** | |
| --- | --- | --- | --- |
| **Premise 1** | All men are mortal. | **Premise 1** | The sun rose this morning. |
| **Premise 2** | Socrates is a man. | **Premise 2** | The sun rose yesterday morning. |
| **Conclusion** | Socrates is mortal. | **Premise 3** | The sun rose two days ago. |
| | | **Premise 4** | The sun rose three days ago. |
| | | **Conclusion** | The sun will rise tomorrow. |

An inductive argument cannot be proved beyond a shadow of a doubt. For example, it's a pretty good bet that the sun will come up tomorrow, but the sun not coming up presents no logical contradiction.

On the other hand, a deductive argument can have logical certainty, but it must be properly constructed. Consider the examples below.

| **True Conclusion from an Invalid Argument** | **False Conclusion from a Valid Argument** |
| --- | --- |
| All men are mortal.<br>Socrates is mortal.<br>Therefore Socrates is a man. | All astronauts are men.<br>Julia Roberts is an astronaut.<br>Therefore, Julia Roberts is a man. |
| Even though the above conclusion is true, the argument is based on invalid logic. Both men and women are mortal. Therefore, Socrates could be a woman. | In this case, the conclusion is false because the premises are false. However, the logic of the argument is valid because *if* the premises were true, then the conclusion would be true. |

A **counterexample** is an example given in which the statement is true but the conclusion is false when we have assumed it to be true. If we said "All cocker spaniels have blonde hair," then a counterexample would be a red-haired cocker spaniel. If we made the statement, "If a number is greater than 10, it is less than 20," we can easily think of a counterexample, like 35.

**Example 1:**   Which argument is valid?

If you speed on Hill Street, you will get a ticket.
If you get get a ticket, you will pay a fine.

(A) I paid a fine, so I was speeding on Hill Street.
(B) I got a ticket, so I was speeding on Hill Street.
(C) I exceeded the speed limit on Hill Street, so I paid a fine.
(D) I did not speed on Hill Street, so I did not pay a fine.

**Solution:**   C is valid.
A is incorrect. I could have paid a fine for another violation.
B is incorrect. I could have gotten a ticket for some other violation.
D is incorrect. I could have paid a fine for speeding somewhere else.

**Example 2:**   Assume the given proposition is true. Then, determine if each statement is true or false.

Given: If a dog is thirsty, he will drink.

| | | |
|---|---|---|
| (A) | If a dog drinks, then he is thirsty. | T or F |
| (B) | If a dog is not thirsty, he will not drink. | T or F |
| (C) | If a dog will not drink, he is not thirsty. | T or F |

**Solution:**   A is false. He is not necessarily thirsty; he could just drink because other dogs are drinking or drink to show others his control of the water. This statement is the **converse** of the original. The converse of the statement "If A, then B" is "If B, then A."

B is false. The reasoning from A applies. This statement is the **inverse** of the original. The inverse of the statement "If A, then B" is "If not A, then not B."

C is true. It is the **contrapositive**, or the complete opposite of the original. The contrapositive says "If not B, then not A."

**For numbers 1–4, assume the given proposition is true. Then, determine if the statements following it are true or false.**

All squares are rectangles.

1. All rectangles are squares.   T or F
2. All non-squares are non-rectangles.   T or F
3. No squares are non-rectangles.   T or F
4. All non-rectangles are non-squares.   T or F

# 12.3   Logic

A **conditional statement** is a type of logical statement that has two parts, a **hypothesis** and a **conclusion**. The statement is written in "if-then" form, where the "if" part contains the hypothesis and the "then" part contains the conclusion. For example, let's start with the statement "Two lines intersect at exactly one point." We can rewrite this as a conditional statement in "if-then" form as follows:

$$\underbrace{\text{If two lines intersect}}_{\text{hypothesis}}, \text{ then } \underbrace{\text{their intersection is at exactly one point}}_{\text{conclusion}}.$$

Conditional statements may be true or false. To show that a statement is false, you need only to provide a single **counterexample** which shows that the statement is not always true. To show that a statement is true, on the other hand, you must show that the conclusion is true for all occasions in which the hypothesis occurs. This is often much more difficult.

**Example 3:**     Provide a counterexample to show that the following conditional statement is false:
If $x^2 = 4$, then $x = 2$.

To begin with, let $x = -2$.
The hypothesis is true, because $(-2)^2 = 4$.
For $x = -2$, however, the conclusion is false even though the hypothesis is true. Therefore, we have provided a counterexample to show that the conditional statement is false.

The **converse** of a conditional statement is an "if-then" statement written by switching the hypothesis and the conclusion. For example, for the conditional statement "If a figure is a quadrilateral, then it is a rectangle," the converse is "If a figure is a rectangle, then it is a quadrilateral."

The **inverse** of a conditional statement is written by negating the hypothesis and conclusion of the original "if-then" conditional statement. Negating means to change the meaning so it is the negative, or opposite, of its original meaning. The inverse of the conditional statement "If a figure is a quadrilateral, then it is a rectangle" is "If a figure is **not** a quadrilateral, then it is **not** a rectangle."

The **contrapositive** of a conditional statement is written by negating the converse. That is, switch the hypothesis and conclusion of the original statement, and make them both negative. The contrapositive of the conditional statement "If a figure is a quadrilateral, then it is a rectangle" is "If a figure is not a rectangle, then it is not a quadrilateral."

**Example 4:**   Given the conditional statement "If $m\angle F = 60°$, then $\angle F$ is acute. Write the converse, inverse and contrapositive.

**Step 1:**   The converse is constructed by switching the hypothesis and the conclusion: If $\angle F$ is acute, then $m\angle F = 60°$.

**Step 2:**   The inverse is constructed by negating the original statement: If $m\angle F \neq 60°$, then $\angle F$ is not acute.

**Step 3:**   The contrapositive is the negation of the converse: If $\angle F$ is not acute, then $m\angle F \neq 60°$.

**Answer the following problems about geometry logic.**

1. Rewrite the following as a conditional statement in "if-then" form: A number divisible by 8 is also divisible by 4.

2. Write the converse of the following conditional statement: If two circles have equal radii, then the circles are congruent.

3. Given the conditional statement: If $x^4 = 81$, then $x = 3$. Is the statement true? Provide a counterexample if it is false.

4. Given the statement: A line contains at least two points. Write as a conditional statement in "if-then" form, then write the converse, inverse, and contrapositive of the conditional statement.

5. "If a parallelogram has four congruent sides, then it is a rhombus." Write the converse, inverse, and contrapositive for the conditional statement. Which are true? Which are false?

6. "If a triangle has one right angle, then the acute angles are complementary." Write the converse, inverse, and contrapositive for the conditional statement. Indicate whether each is true or false. Can all the statements be either true or false? Explain.

7. "If a rectangle has four congruent sides, then it is a square." Write the contrapositive for the conditional statement and indicate whether it is true or false. Give a counterexample if it is false.

8. Show why a conditional statement and its inverse are always logically equivalent. Similarly, show why a statement's converse and inverse are logically equivalent.

## 12.4   Using Diagrams to Solve Problems

Problems that require logical reasoning cannot always be solved with a set formula. Sometimes, drawing diagrams can help you see the solution.

**Example 5:**    Yvette, Barbara, Patty, and Nicole agreed to meet at the movie theater around 7:00 p.m. Nicole arrived before Yvette. Barbara arrived after Yvette. Patty arrived before Barbara but after Yvette. What is the order of their arrival?

| Nicole | Yvette | Patty | Barbara |
|--------|--------|-------|---------|
| 1st | 2nd | 3rd | 4th |

**Use a diagram to answer each of the following questions.**

1. Javy, Thomas, Pat, and Keith raced their bikes across the playground. Keith beat Thomas but lost to Pat and Javy. Pat beat Javy. Who won the race?

2. Jeff, Greg, Pedro, Lisa, Macy, and Kay eat lunch together at a round table. Kay wants to sit beside Pedro, Pedro wants to sit next to Lisa, Greg wants to sit next to Macy, and Jeff wants to sit beside Kay. Macy would rather not sit beside Lisa. Which two people should sit on each side of Jeff?

3. Three teams play a round-robin tournament where each team plays every other team. Team A beat Team C. Team B beat Team A. Team B beat Team C. Which team is the best?

4. Caleb, Thomas, Ginger, Alex, and Janice are in the lunch line. Thomas is behind Alex. Caleb is in front of Alex but behind Ginger and Janice. Janice is between Ginger and Caleb. Who is third in line?

5. Ray, Fleta, Paula, Joan, and Henry hold hands to make a circle. Joan is between Ray and Paula. Fleta is holding Ray's other hand. Paula is also holding Henry's hand. Who must be holding Henry's other hand?

6. The Bears, the Cavaliers, the Knights, and the Lions all competed in a track meet. One team from each school ran the 400-meter relay race. The Bears beat the Knights but lost to the Cavaliers. The Lions beat the Cavaliers. Who finished first, second, third, and fourth?

## 12.5   Trial and Error Problems

Sometimes problems can only be solved by trial and error. You have to guess at a solution, and then check to see if it will satisfy the problem. If it does not, you must guess again until you get the right answer.

**Solve the following problems by trial and error. Make a chart of your attempts so that you don't repeat the same attempt twice.**

1. Becca had 5 coins consisting of one or more quarters, dimes, and nickels that totaled $0.75. How many quarters, dimes, and nickels did she have?

2. Ryan needs to buy 42 cans of soda for a party at his house. He can get a six pack for $1.80, a box of 12 for $3.00, or a case of 24 for $4.90. What is the least amount of money Ryan must spend to purchase the 42 cans of soda?

3. Jana had 10 building blocks that were numbered 1 to 10. She took three of the blocks and added up the three numbers to get 27. Which three blocks did she pick?

4. Hank had 10 coins. He had 3 quarters, 3 dimes, and 4 nickels. He bought a candy bar for 75¢. How many different ways could he spend his coins to pay for the candy bar?

5. Refer to question 4. If Hank used 6 coins to pay for the candy bar, how many of his quarters did he spend?

6. The junior varsity basketball team needs to order 38 pairs of socks for the season. The coach can order 1 pair for 2.45, 6 pairs for $12.95, or 10 pairs for $20.95. What is the least amount of money he will need to spend to purchase exactly 38 pairs of socks?

7. Tyler has 5 quarters, 10 dimes, and 15 nickels in change. He wants to buy a notebook for $2.35 using the change that he has. If he wants to use as many of the coins as possible, how many quarters will he spend?

8. Kevin is packing up his room to move to another city. He has the following items left to pack.

| | |
|---|---|
| comic book collection | 7 pounds |
| track trophy | 3 pounds |
| coin collection | 13 pounds |
| soccer ball | 1 pound |
| model car | 6 pounds |

If Kevin has a large box that will hold 25 pounds, what items should he pack in it to get the most weight without going over the box's weight limit?

## 12.6   Pattern Problems

Some problems follow a pattern. You must read these problems carefully and recognize the pattern.

**Example 6:**   Jason wants to swim in the ocean from the shore to the end of a pier 38 feet out. He must swim against the ocean's waves. For every 10 feet he swims forward, a wave takes him back 3 feet. How many total feet will he swim to touch the end of the pier?

**Step 1:**   Draw a diagram or create a table to help you visualize the problem.

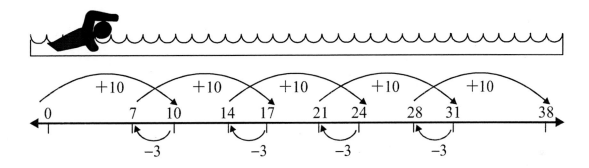

**Step 2:**   Determine how many feet he swam by going forward 10 feet at a time. Look at the diagram above. There are 5 forward arrows that each represent 10 feet to get Jason to the 38 feet mark. $10 \times 5 = 50$ feet
(Since Jason reached the pier before the last wave pushed him back, we do not subtract 3 feet for the last wave.)

**Read each of the following questions carefully. Draw a diagram or create a table to help you visualize the pattern. Then answer the question.**

1. In a strong-man contest, the contestants must pull a car up an incline 13 feet long. Joe is the first contestant. With every tug, Joe pulls the car up the incline 3 feet. Before he can tug again, the car rolls down 1 foot. How many tugs will it take Joe to get the car up the incline?

2. Greta goes on a jungle safari and steps in quicksand. She sinks 25 inches. Her partner loops a rope around a tree limb directly above her so she can pull herself out. For every 5 inches she is able to pull up, she slips back down 1 inch. How many total inches must she climb up to completely free herself from the 25 inches of quicksand?

3. Randal is starting a job as a commissioned salesman. He establishes 5 clients the first week of his job. His goal is to double his clientele every week. If he can accomplish his goal, how many weeks will it take him to establish 320 clients?

4. A five-year-old decides to run the opposite way on the moving sidewalk at the airport. For every 7 feet he runs forward, the sidewalk moves him 4 feet backward. How many total feet will he run to get to the end of a sidewalk that is 19 feet long?

5. A certain bacteria culture doubles its population every minute. Every 3 minutes, the entire culture decreases by half. If a culture starts with a population of 1000, what would be the population of the bacteria at 6 minutes? Hint: The population equals 1000 at 0 minutes.

6. Rita's mom needs to bake 5 dozen chocolate chip cookies for a bake sale. For every dozen cookies she bakes, her children eat two. How many dozen cookies will she have to bake to have 5 dozen for the bake sale?

7. A retail store plans on building 3 new stores each year. They expect that 1 store will go out of business every 21 years. How many years would it take to establish 26 stores?

8. Jasmine and her family sold Girl Scout cookies this spring at the mall. After every seven boxes she sold, her family bought a box to celebrate. Jasmine sold a total of 64 boxes. How many of those boxes did her family buy?

9. A pitcher plant has leaves modified as pitchers for trapping and digesting insects. An insect is lured to the edge of a 7-inch leaf and slips down 3 inches. For every 3 inches it slips down, it manages to climb up 2 inches. It then starts slipping back down 3 inches towards the digestive juices at the bottom of the leaf. How many total inches will the insect slip down until it reaches the bottom?

# 12.7   Reasonable Solutions

In the real world, estimates can be very useful. The best approach to finding estimates is to round off all numbers in the problem. Then solve the problem, and choose the closest answer. If money problems have both dollars and cents, round to the nearest dollar or ten dollars. $44.86 rounds to $40.

**Example 5:**   Which is a reasonable answer? $1580 \div 21$

 A. 80     B. 800     C. 880     D. 8000

**Step 1:**   Round off the numbers in the problem. 1580 rounds to 1600   21 rounds to 20

**Step 2:**   Work the problem. $1600 \div 20 = 80$     The closest answer is A. 80.

**Choose the best answer below.**

1. Which is a reasonable answer? $544 \times 12$
   (A) 54
   (B) 500
   (C) 540
   (D) 5400

2. Jeff buys a pair of pants for $45.95, a belt for $12.97, and a dress shirt for $24.87. Estimate about how much he spends.
   (A) $60
   (B) $70
   (C) $80
   (D) $100

3. For lunch, Marcia eats a sandwich with 187 calories, a glass of skim milk with 121 calories, and 2 brownies with 102 calories each. About how many calories does she consume?
   (A) 300
   (B) 350
   (C) 480
   (D) 510

4. Which is a reasonable answer? $89,900 \div 28$
   (A) 300
   (B) 500
   (C) 1000
   (D) 3000

5. Which is a reasonable answer? $74,295 - 62,304$

(A) $12,000$

(B) $11,000$

(C) $10,000$

(D) $1000$

6. Delia buys 4 cans of soup at $0.99 each, a box of cereal for $4.78, and 2 frozen dinners at $3.89 each. About how much does she spend?

(A) $10.00

(B) $11.00

(C) $13.00

(D) $17.00

7. Which is the best estimate? $22,480 + 5516$

(A) 2800

(B) $17,000$

(C) $28,000$

(D) $32,000$

8. Which is the best estimate? $23,895 \div 599$

(A) $20.00

(B) $30.00

(C) $40.00

(D) $50.00

9. Tracy needs a pack of paper, 2 folders, a protractor, and 6 pencils. Using the chart below, about how much money does she need?

(A) $3.00

(B) $4.00

(C) $5.00

(D) $6.00

10. Jake needs 2 pencils, 3 erasers, a binder, and a compass. About how much money will he need according to the chart below?

(A) $6.00

(B) $7.00

(C) $8.00

(D) $9.00

**School Store Price List**

| Pencils | Erasers | Folders | Binders | Compass | Protractor | Paper | Pens |
|---|---|---|---|---|---|---|---|
| 2 for $0.78 | $0.59 | $0.21 | $2.79 | $1.59 | $0.89 | $1.29 | $1.10 |

# Chapter 12 Review

**For numbers 1–4, assume the given proposition is true. Then determine if the statements following it are true or false.**

All whales are mammals.

1. All non-whales are non-mammals.     T or F

2. If a mammal lives in the sea, it is a whale.     T or F

3. All mammals are whales.     T or F

4. All non-mammals are non-whales.     T or F

**For numbers 5–8, determine whether the situation is showing deductive or inductive logic.**

5. A group of students were given three descriptions about a person's job. They were then told to decide what type of job title the person has.

6. When traveling in a car on a family vacation, I noticed that I could see the ocean to my left and palm trees to my right. I concluded that my family and I were going to the beach.

7. Sammy asked her friend, Amy, to give her a good reason to get a summer job. Amy gave Sammy four good reasons to get a job.

8. The neighbor's cars are in the driveway and all of the lights in the house are off, so they must be sleeping.

**Look at statements 9–12. Determine whether the statements are true always, sometimes, or never.**

9. Quadratic equations have two solutions.

10. If you graph a linear equation, the graph will be a straight line.

11. When multiplying both sides of an inequality by a number, you must reverse the direction of the inequality symbol.

12. When you take the absolute value of a number, you are making the number negative.

**Solve the following problems.**

13. "If a triangle is isosceles, then its base angles are congruent." Write the contrapositive for the conditional statement. Is the statement true or false? Is the contrapositive true or false? If false, give a counterexample.

14. "If the radius of a circle is doubled, then the area of the circle is increased by a factor of four." Write the converse, inverse, and contrapositive for the conditional statement. Indicate which ones are true or false.

15. "If today is Tuesday, then it is raining." Write the converse, inverse, and contrapositive for the conditional statement. Could the statements be true? Give a counterexample to prove each statement false.

16. Andrea spent $1.24 for toothpaste. She had quarters, dimes, nickels, and pennies. How many dimes did she use if she used a total of 14 coins?

17. Cody spent $0.59 on a hotdog. He had quarters, dimes, nickels, and pennies in his pocket. He gave the cashier 9 coins for exact payment. How many quarters did he give the cashier?

18. Vince, Hal, Weng, and Carl raced on roller blades down a hill. Vince beat Carl. Hal finished before Vince but after Weng. Who won the race?

19. A veterinarian's office has a weight scale in the lobby to weigh pets. The scale will weigh up to 250 pounds. The following dogs are in the lobby:

| | |
|---|---|
| Pepper | 23 pounds |
| Jack | 75 pounds |
| Trooper | 45 pounds |
| Precious | 25 pounds |
| Coco | 120 pounds |

Which dogs could you put on the scale to get as close to 250 pounds as possible, without going over? How much would they weigh?

20. Felix has set a goal to increase his running speed by 1 minute per mile every 5 weeks. He starts out running 1 mile in 11 minutes. If he can accomplish his goal, how many weeks will it take him to run a mile in 8 minutes?

21. Mr. Sanders is hiring someone for the newest position that has opened up at his company. He interviewed four people for the job: Rick, Luca, Janelle, and Jacob. To help with his decision, he puts the four candidates in order from best qualified to least qualified. Janelle and Rick are better qualified than Luca. Rick and Luca are better qualified than Jacob, but Luca is less qualified than Janelle. If Janelle is not as qualified for the position as Rick, who is Mr. Sanders going to hire for the new opening in his company?

# Chapter 12 Test

**For 1–3, chose which argument is valid.**

1. If I oversleep, I miss breakfast. If I miss breakfast, I cannot concentrate in class. If I do not concentrate in class, I make bad grades.

   A I made bad grades today, so I missed breakfast.

   B I made good grades today, so I got up on time.

   C I could not concentrate in class today, so I overslept.

   D I had no breakfast today, so I overslept.

2. If I do not maintain my car regularly, it will develop problems. If my car develops problems, it will not be safe to drive. If my car is not safe to drive, I cannot take a trip in it.

   A If my car develops problems, I did not maintain it regularly.

   B I took a trip in my car, so I maintained it regularly.

   C If I maintain my car regularly, it will not develop problems.

   D If my car is safe to drive, it will not develop problems.

3. If two triangles have all corresponding sides and all corresponding angles congruent, then they are congruent triangles. If two triangles are congruent, then they are similar triangles.

   A Similar triangles have all sides and all angles congruent.

   B If two triangles are similar, then they are congruent.

   C If two triangles are not congruent, then they are not similar.

   D If two triangles have all corresponding sides and angles congruent, then they are similar triangles.

4. Cynthia is asked to list five duties of the President. What type of logic is Cynthia using?

   A mathematical reasoning

   B inductive reasoning

   C intuitive reasoning

   D deductive reasoning

5. You can find the equation of a line by using two points that lie on that line. When is this statement true?

   A always

   B sometimes

   C never

   D cannot be determined

6. When using order of operations to simplify an algebraic expression, the first step is to get rid of the exponents. When is this statement true?

   A always

   B sometimes

   C never

   D cannot be determined

7. Which is a reasonable estimate of the answer?

   $16,729 \div 100$

   A 16,700

   B 1,670

   C 167

   D 17

# End-of-Course Assessment Algebra I Reference Sheet

### Pythagorean Theorem

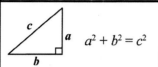

$$a^2 + b^2 = c^2$$

### Distance Formula

$$d = \sqrt{(x_2 - x_1)^2 + (y_2 - y_1)^2}$$

$d$ = distance between points 1 and 2

### Midpoint Formula

$$M = \left( \frac{x_1 + x_2}{2}, \ \frac{y_1 + y_2}{2} \right)$$

$M$ = point halfway between points 1 and 2

### Standard Form of a Linear Equation

$$Ax + By = C$$

(where $A$ and $B$ are not both zero)

### Standard Form of a Quadratic Equation

$$ax^2 + bx + c = 0$$

(where $a \neq 0$)

### Quadratic Formula

$$x = \frac{-b \pm \sqrt{b^2 - 4ac}}{2a}$$

(where $ax^2 + bx + c = 0$ and $a \neq 0$)

### Equation of a Line

**Slope-Intercept Form:** $y = mx + b$
where $m$ = slope and $b$ = $y$-intercept

**Point-Slope Form:** $y - y_1 = m(x - x_1)$

### Simple Interest Formula

$$I = prt$$

where $I$ = interest

$p$ = principal

$r$ = rate

$t$ = time

### Slope of a Line

Let $(x^1, y^1)$ and $(x^2, y^2)$ be two points in a plane

$$\text{slope} = \frac{\text{change in } y}{\text{change in } x} = \frac{y^2 - y^1}{x^2 - x^1}$$

(where $x_2 \neq x_1$)

| Shape | Formulas for Area (*A*) and Circumference (*C*) | |
|---|---|---|
| **Triangle** | $A = \frac{1}{2}bh = \frac{1}{2} \times$ base $\times$ height | |
| **Trapezoid** | $A = \frac{1}{2}(b_1 + b_2)h = \frac{1}{2} \times$ sum of bases $\times$ height | |
| **Parallelogram** | $A = bh =$ base $\times$ height | |
| **Circle** | $A = \pi r^2 = \pi \times$ square of radius <br> $C = 2\pi r = 2 \times \pi \times$ radius | $\pi \approx 3.14$ <br> or $\pi \approx \frac{22}{7}$ |
| **Figure** | **Formulas for Volume (*V*) and Surface Area (*SA*)** | |
| **Cube** | $SA = 6s^2 = 6 \times$ length of side squared | |
| **Cylinder (total)** | $SA = 2\pi rh + 2\pi r^2$ <br> $SA = 2 \times \pi \times$ radius $\times$ height $+ 2 \times \pi \times$ radius squared | $\pi \approx 3.14$ <br> or <br> $\pi \approx \frac{22}{7}$ |
| **Sphere** | $SA = 4\pi r^2 = 4 \times \pi \times$ radius squared <br> $V = \frac{4}{3}\pi r^3 = \frac{4}{3} \times \pi \times$ radius cubed | |
| **Cone** | $V = \frac{1}{3}\pi r^2 h = \frac{1}{3} \times \pi \times$ radius squared $\times$ height | |
| **Pyramid** | $V = \frac{1}{3}Bh = \frac{1}{3} \times$ area of base $\times$ height | |
| **Prism** | $V = Bh =$ area of base $\times$ height | |

# Practice Test 1

## Session 1

1. Simplify $\dfrac{x+7}{7x^2-343}$.

    **A** $\dfrac{1}{2}$

    **B** $\dfrac{1}{7\left(x^2-343\right)}$

    **C** $\dfrac{1}{7\left(x+7\right)}$

    **D** $\dfrac{1}{7\left(x-7\right)}$

    A1.7.1

2. Solve the proportion $\dfrac{3x+2}{2}=\dfrac{3x-12}{6}$.

    **A** $x=1$
    **B** $x=-1$
    **C** $x=-3$
    **D** $x=-6$

    A1.7.2

3. Use the substitution method to solve the pair of equations $y=3x+11$ and $y=2x$.

    **A** $(-1,-2)$
    **B** $(1,2)$
    **C** $(-11,-22)$
    **D** $(11,22)$

    A1.5.3

4. Which property is demonstrated below?

    $(a\times b)\times c=a\times(b\times c)$

    **A** Associative Property of Multiplication
    **B** Commutative Property of Multiplication
    **C** Identity Property of Multiplication
    **D** Inverse Property of Multiplication

    A1.1.3

5. Which is the graph of $x-3y=6$?

    **A**

    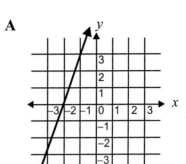

    **B**

    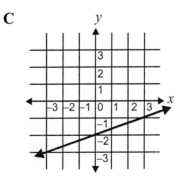

    **C**

    **D**

    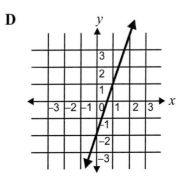

    A1.4.1

187

6. Solve for $z$. $\dfrac{3z + 7}{y} = x$

   **A**   $z = \dfrac{xy - 7}{3}$

   **B**   $z = \dfrac{3x + 7}{y}$

   **C**   $z = \dfrac{x - 7y}{3}$

   **D**   $z = \dfrac{7xy}{3}$

A1.2.2

7. Solve the linear inequality $3x - 16 > 5x + 12$.

   **A**   $x > -12$
   **B**   $x > -14$
   **C**   $x < -7$
   **D**   $x < -14$

A1.2.4

8. Solve the equation $2 - 12a + 5a = a + 6$.

   **A**   $a = -\dfrac{1}{3}$

   **B**   $a = -\dfrac{1}{2}$

   **C**   $a = -\dfrac{1}{4}$

   **D**   $a = -\dfrac{3}{4}$

A1.2.1

9. Is $y = x^3$ a function?

   **A**   Yes, because it passes the horizontal line test.
   **B**   Yes, because it passes the vertical line test.
   **C**   No, because it fails the horizontal line test.
   **D**   No, because it fails the vertical line test.

A1.3.3

10. Solve $x^2 + 7x + 12 = 0$ using the quadratic formula.

   **A**   $x = -4, -3$
   **B**   $x = 2, 5$
   **C**   $x = -3, 10$
   **D**   $x = -1, 8$

A1.8.6

11. Flying at an altitude of 500 meters, Cheryl notices the air temperature outside the plane is 12°C. When the plane climbs to an altitude of 1500 meters, the outside temperature reading is 5°C. Which of these graphs depicts temperature as a linear function of altitude to represent this situation?

**A**

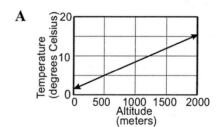

**B**

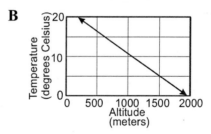

**C**

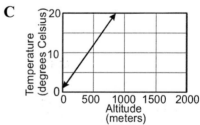

**D**

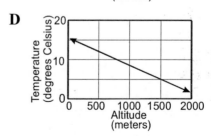

A1.3.1

12. Which numbers from the set $\{-3, -1, 0, 4\}$ make the inequality $11x + 3 \le x^3$ true?

   **A** $\{0, 4\}$
   **B** $\{-3, -1, 0\}$
   **C** $\{-3, -1, 4\}$
   **D** $\{-1, 0, 4\}$

   A1.2.3

13. Solve the combined inequality
   $-13 < \dfrac{3x - 5}{2} \le 3$.

   **A** $-21 < x \le 11$

   **B** $1 \le x < 4$

   **C** $\frac{-7}{3} < x \le 8$

   **D** $-7 < x \le \frac{11}{3}$

   A1.2.5

14. Solve the inequality $\frac{3}{7}x - 4 < 8$

   **A** $x < 28$
   **B** $x < 4$
   **C** $x > -14$
   **D** $x > 7$

   A1.2.4

15. Simplify $\sqrt{300}$.

   **A** $30$
   **B** $10\sqrt{3}$
   **C** $9\sqrt{10}$
   **D** $20\sqrt{15}$

   A1.1.2

16. What is the domain of $x - 2 = y + 6$?

   **A** $-6 \le x \le 2$
   **B** $-6 \le x < \infty$
   **C** $-\infty < x \le 2$
   **D** $-\infty < x < \infty$

   A1.3.4

17. Which system of inequalities defines the shaded region below?

   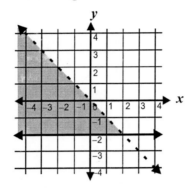

   **A** $y < -1, y \ge -2x$
   **B** $y < -x, y \ge -2$
   **C** $y > -2x, y \le -1$
   **D** $y < -x, y \ge -2x$

   A1.5.2

18. Use substitution to find a common solution for the pair of equations $5x + 2y = 1$ and $2x + 4y = 10$.

   **A** $(1, -2)$
   **B** $(3, 1)$
   **C** $(1, 2)$
   **D** $(-1, 3)$

   A1.5.3

19. Use multiplication with the subtraction method to solve the equations: $2x - y = 2$ and $4x - 9y = -3$.

   **A** $(1.5, 1)$
   **B** $(2, 2)$
   **C** $(-3, -1)$
   **D** $(0, \frac{1}{3})$

   A1.5.5

20. Simplify:
   $2(15x^2 + x - 7) - 5(6x^2 - x - 14)$

   **A** $7x + 56$
   **B** $-60x^2 + 3x + 84$
   **C** $60x^2 - 7x - 28$
   **D** $15x^2 + 11x$

   A1.6.1

21. Find the greatest common factor in the following terms:
$11x^5y + 5x^4y^2 - 2x^3y^3 + 6x^2 - 3xy^5$.

A $y$

B $x$

C $x^2y$

D $xy^3$

A1.6.6

22. Devin is practicing golf at the driving range. The equation that represents the height of his ball is $-0.5t^2 + 12t = s$, where $s$ is the number of feet and $t$ is the number of seconds. If the ball is at 31.5 ft in the air, how many seconds have gone by?

A 5 seconds

B 3 seconds

C 21 seconds

D 3 seconds or 21 seconds

A1.8.7

23. You are selling candy bars for $0.85 a piece, and you need to raise at least $135.00. Write an inequality that shows how many candy bars you need to sell.

A1.2.6

24. Solve $(x + 5)^2 = 49$

A1.8.3

25. Solve $\sqrt{-3x + 18} = x$

A1.8.8

26. Which numbers in the set $\{-6, -1, 3, 6, 11\}$ make the inequality $-2x < 1$ true?

A1.2.3

27. Convert 20 mg/L to grams per liter.

A1.1.5

28. Sketch the graph of $x^3 + x - 2$.

A1.8.1

# Session 2

1. What is the square root of $x^3y^8z$?

  **A**  $xy^4z$
  **B**  $xy^4\sqrt{xz}$
  **C**  $y^4\sqrt{x^3z}$
  **D**  $z\sqrt{x^3y^8}$

A1.6.3

2. Divide $a^3b^6c$ by $bc$.

  **A**  $a^3b^5c$
  **B**  $a^3b^5$
  **C**  $b^5$
  **D**  $ab^3$

A1.6.2

3. What is the domain of $y = 3\sqrt{x}$?

  **A**  $0 \leq x < \infty$
  **B**  $0 \leq y < \infty$
  **C**  $-\infty < x \leq 0$
  **D**  $-\infty < x < \infty$

A1.3.4

4. Is $y = x^2 - 5$ a function?

  **A**  Yes, because it passes the vertical line test.
  **B**  Yes, because it passes the horizontal line test.
  **C**  No, because it fails the vertical line test.
  **D**  No, because it fails the horizontal line test.

A1.3.3

5. Is $y = -x + 2$ a function?

  **A**  Yes, because it passes vertical line test.
  **B**  Yes, because it fails horizontal line test.
  **C**  No, because it passes horizontal line test.
  **D**  No, because it fails the vertical line test.

A1.3.4

6. What is the $x$-intercept of the equation $y = 2x + \frac{3}{2}$?

  **A**  $\left(-\frac{3}{4}, 0\right)$
  **B**  $\left(\frac{3}{2}, 0\right)$
  **C**  $(-1, 0)$
  **D**  $\left(\frac{5}{7}, 0\right)$

A1.4.2

7. What is the following equation in slope-intercept form?
$$y - 3 = 2(x - 2)$$

  **A**  $y = x - 6$
  **B**  $y = x + 3$
  **C**  $y = 2x - 1$
  **D**  $y = 2x - 4$

A1.4.3

8. A line goes through the points $(-5, 3)$ and $(0, 13)$. What is the equation of the line that is perpendicular to the first line and passes through the point $(-2, 9)$?

  **A**  $y = x + 8$
  **B**  $y = x + 3$
  **C**  $y = -\frac{1}{2}x + 8$
  **D**  $y = -2x + 13$

A1.4.4

9. There are 2 less than 3 times as many dimes as there are quarters. The change totals to $6.95. How many quarters are there?

  **A**  13
  **B**  9
  **C**  17
  **D**  11

A1.5.6

10. Solve the pair of equations $6x + 2y = 8$ and $2y - 2x = 14$ using substitution.

**A** $(1, 1)$

**B** $(1, 8)$

**C** $\left(-\frac{3}{4}, \frac{25}{4}\right)$

**D** $(-2, 10)$

A1.5.3

11. Solve the pair of equations $2x - y = 2$ and $4x - 9y = -3$.

**A** $\left(\frac{3}{2}, 8\right)$

**B** $\left(\frac{3}{2}, 1\right)$

**C** $(3, 4)$

**D** $(6, 3)$

A1.5.5

12. Solve $3x^2 + 12x + 12 = 0$ using the quadratic formula.

**A** $x = -2$

**B** $x = -3, 4$

**C** $x = 5, 7$

**D** $x = 6$

A1.8.6

13. Solve $x^2 + 3x - 28 = 0$ by factoring.

**A** $x = -7, 4$

**B** $x = 1, 2,$

**C** $x = -5, 8$

**D** $x = 0, 3$

A1.8.2

14. What is the range of $y = x^2 - 1$?

**A** $-1 \leq y < \infty$

**B** $0 \leq y < \infty$

**C** $-\infty < x < \infty$

**D** $-1 \leq x < \infty$

A1.3.4

15. Line $MN$ goes through points $(-2, -4)$ and $(2, 2)$. Line $OP$ is perpendicular to line $MN$ and passes through point $(0, -1)$. What is the equation of line $OP$?

**A** $y = -\frac{2}{3}x - 1$

**B** $y = \frac{3}{2}x - 1$

**C** $y = 2x - 1$

**D** $y = \frac{1}{2}x - 1$

A1.4.4

16. Which system of inequalities defines the shaded region below?

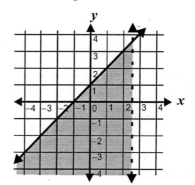

**A** $y \leq x + 1$
   $x < 2.5$

**B** $y \geq x + 1$
   $x > 2.5$

**C** $y \leq x + 1$
   $x > 2.5x$

**D** $y > x + 1$
   $x < 2.5y$

A1.5.2

17. Simplify the ratio $\dfrac{x^2 - 4}{x^2 + 9x + 14}$.

**A** $\dfrac{x - 2}{x + 2}$

**B** $\dfrac{x - 2}{x - 7}$

**C** $\dfrac{x + 2}{x + 7}$

**D** $\dfrac{x - 2}{x + 7}$

A1.7.1

18. Solve the equation $\sqrt{-10x - 1} = 3x$.

  **A** $x = -\frac{1}{10}, 0$

  **B** $x = -1, -\frac{1}{9}$

  **C** $x = \frac{1}{3}, 3$

  **D** $x = -\frac{1}{9}, -\frac{1}{10}$

  A1.8.8

19. Line $WX$ passes through points $(3, -1)$ and $(0, 1)$. If line $YZ$ is parallel to $WX$ and passes through $(-6, 2)$, what is the equation of line $YZ$?

  **A** $y = \frac{3}{2}x + 11$

  **B** $y = -\frac{2}{3}x - 2$

  **C** $y = -\frac{3}{2}x - 7$

  **D** $y = \frac{2}{3}x + 6$

  A1.4.4

20. Solve $x^2 + 6x - 7 = 0$ using a calculator.

  **A** $x = -7, 1$
  **B** $x = -1, 7$
  **C** $x = 1, 7$
  **D** $x = -1, -7$

  A1.8.9

21. Solve the proportion $\dfrac{5x + 4}{6} = \dfrac{x + 8}{3}$.

  A1.7.2

22. Use addition to solve the equations:
    $3x - y = 2$ and $5x + y = 6$.

  A1.5.4

23. Factor the equation $3x^2 - 6x - 189$.

  A1.6.7

24. Is $y = x^2 - 3x + 7$ function? Why or why not? If it is, what is the domain and range of the equation?

  A1.3.3

25. Martina, Marc, and Viktor ran 3 miles each during their gym class today. At the end of their run, the three of them drank a total of 138 ounces of water. Viktor had 5 more ounces of water than Marc. Martina had 13 ounces more than half as much as Marc. How many ounces of water did Martina drink?

  A1.2.6

26. Which numbers in the set $\{-7, -5, -3, -1, 0, 2, 4, 5\}$ make the inequality $(-x)^2 - 3 > 7$ true?

  A1.2.3

# Practice Test 2

## Session 1

1. Use addition to solve the equations:
$x - y = 1$ and $x + y = 6$.

  **A** $\left(\frac{7}{4}, \frac{3}{4}\right)$

  **B** $\left(\frac{10}{3}, \frac{8}{3}\right)$

  **C** $\left(\frac{5}{2}, \frac{7}{2}\right)$

  **D** $\left(\frac{7}{2}, \frac{5}{2}\right)$

A1.5.4

2. Solve the algebraic proportion
$\dfrac{x + 2}{7} = \dfrac{x + 4}{21}$.

  **A** $x = -1$

  **B** $x = 2$

  **C** $x = -4$

  **D** $x = 7$

A1.7.2

3. What is the domain of $y = x^3 - 7$?

  **A** $-7 \leq x < \infty$

  **B** $-\infty < x < \infty$

  **C** $-\infty < y \leq -7$

  **D** $-7 \leq y < \infty$

A1.3.4

4. What is the range of the function
$y = -x^2 + 1$?

  **A** $-\infty < y < \infty$

  **B** $-\infty < y \leq 0$

  **C** $-\infty < y \leq 1$

  **D** $1 \leq y < \infty$

A1.3.4

5. Is the graph below a function?

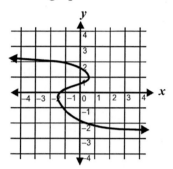

  **A** Yes, because it passes the vertical line test.

  **B** Yes, because it passes the horizontal line test.

  **C** No, because it fails the vertical line test.

  **D** No, because it fails the horizontal line test.

A1.3.3

6. What is 790 centimeters per second in kilometers per hour?

  **A** 28.44 km/hr

  **B** 28, 440 km/hr

  **C** 0.2844 km/hr

  **D** 0.0079

A1.1.5

7. What is the slope of a line represented by the equation $x = 2y - 3$?

  **A** $-2$

  **B** $-\frac{1}{2}$

  **C** $\frac{1}{2}$

  **D** $2$

A1.4.2

8. Which property is demonstrated below?

$a \times b = b \times a$

A Identity Property of Multiplication
B Inverse Property of Multiplication
C Commutative Property of Multiplication
D Associative Property of Multiplication

A1.1.3

9. Solve $3x = \sqrt{39x + 30}$

A $x = 0, 5$

B $x = \frac{3}{2}, 6$

C $x = -\frac{3}{2}, 13$

D $x = -\frac{2}{3}, 5$

A1.8.8

10. Solve $x^2 + 12x + 27 = 0$ by completing the square.

A $x = -9, -3$
B $x = -21, 9$
C $x = -3, 9$
D $x = -21, 3$

A1.8.4

11. Solve $x^2 - 4x - 12 = 0$ using the quadratic formula.

A $x = -2, 2$
B $x = -12, 1$
C $x = -2, 6$
D $x = -5, 7$

A1.8.6

12. Solve the combined inequality
$5x + 3 > 3x - 3 \geq 5x - 13$.

A $3 > x \geq -5$
B $-12 \leq x < 4$
C $6 > x \geq -10$
D $-3 < x \leq 5$

A1.2.5

13. Given the equation $\dfrac{2a - 3b}{c} = 5$, solve for $a$.

A $a = \dfrac{2 + b}{15c}$

B $a = \dfrac{5c + 3b}{2}$

C $b = \dfrac{2a - 5c}{3}$

D $c = \dfrac{2a - 3b}{5}$

A1.2.2

14. Dan is saving money to make a down payment of $600 for a car. In week 1, he saved $219, and in week 2, he saved $183. Which inequality shows how much money Dan needs to save in week 3 to be able to make his down payment?

A $x < 201$
B $x \leq 201$
C $x > 198$
D $x \geq 198$

A1.2.6

15. Simplify the algebraic ratio:

$$\frac{x^5 - 7x^4}{x^5 - 9x^4 + 14x^3} + \frac{3x^2 + 6x}{x^3 - 4x}$$

A $\dfrac{x + 2}{x - 7}$

B $\dfrac{x - 7}{x(x - 2)}$

C $\dfrac{x + 3}{x - 2}$

D $\dfrac{x - 7}{x - 2}$

A1.7.1

16. Find the square root of $a^4 b^{10}$.

A1.6.3

17. Solve for $x$ in the algebraic proportion $7x^2 - x + 3 = 4x + 3$.

   **A** $x = 0, \frac{5}{7}$
   **B** $x = -1, 0$
   **C** $x = \frac{1}{2}, \frac{7}{5}$
   **D** $x = -\frac{5}{7}, 0$

A1.7.2

18. Which set of inequalities corresponds to the shaded region below?

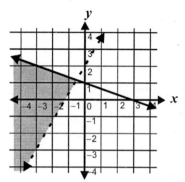

   **A** $y < x, x \leq 1$

   **B** $y > x + 2, y \geq 2x + 3$

   **C** $y \leq \frac{2}{3}x - 2, x > \frac{1}{3}$

   **D** $y \leq -\frac{1}{3}x + 1, y > \frac{5}{3}x + 2$

A1.5.2

19. Which equation below is written in slope-intercept form?

   **A** $3x + 2y = 8$
   **B** $3x + 2y - 8 = 0$
   **C** $y = 4x - 8$
   **D** $4x - 8 = 0$

A1.4.3

20. Factor $18x^2 - 32$.

   **A** $2(3x - 4)(3x + 4)$
   **B** $(6x - 2)(3x + 2)$
   **C** $(9x + 3)(2x + 1)$
   **D** $4(4x^2 - 8)$

A1.6.7

21. Which graph below represents the inequality $x \leq 3$?

   **A**

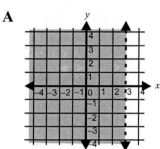

   **B**

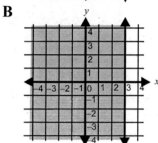

   **C**

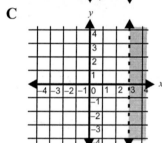

   **D**

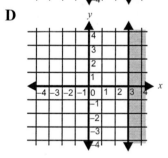

A1.4.6

22. Andrea is going to the grocery store to buy apples and lemons. 13 apples and 15 lemons will cost her $10.44. 3 apples and 7 lemons will cost her $2.94. What is the price of an apple?

   **A** $5.43
   **B** $0.15
   **C** $0.44
   **D** $0.63

A1.5.6

23. Which members of the set

$$\{-3, -2, -1, 0, 1, 2, 3\}$$

are solutions for $-2x + 5 > 10$?

**A** $\{-3, -2\}$
**B** $\{0, 1, 2, 3\}$
**C** $\{-3, -2, -1\}$
**D** $\{-3\}$

A1.2.3

24. Cyrus is testing out his new bottle rocket. The equation that represents the height of the rocket is $s = -2x^2 + 17x$, where $s$ is height in meters and $t$ is time elapsed in seconds. What is the maximum height the rocket will go?

**A** 39.775 m
**B** 36.125 m
**C** 33.545 m
**D** 41.125 m

A1.8.7

25. Line $l$ passes through points $(-6, 1)$ and $(-3, 6)$. Line $m$ is parallel to line $l$ and passes through point $(15, -1)$. What is the equation of line $m$?

**A** $y = \frac{5}{3}x + 11$

**B** $y = \frac{5}{3}x - 26$

**C** $y = -\frac{5}{3}x + 4$

**D** $y = -\frac{5}{3}x - 7$

A1.4.4

26. Which of the following is the graph of the equation $y = x + 2$?

**A**

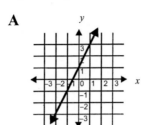

**B**

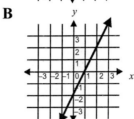

**C**

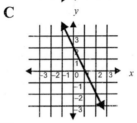

**D**

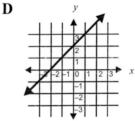

A1.4.1

27. Solve the inequality $-6x + 3 \le 2x - 13$.

A1.2.4

28. Solve the inequality $\dfrac{-2x - 1}{3} < \dfrac{-x - 5}{2}$ and explain each step.

A1.7.2

# Session 2

1. Solve $16x + 12 = 9x + 33$.

   **A** $x = \dfrac{21}{25}$

   **B** $x = 3$

   **C** $x = 6$

   **D** $x = 7$

   A1.2.1

2. If $7r + 5s = 35$, then $r =$

   **A** 5

   **B** 7

   **C** $\dfrac{35 - 5s}{7}$

   **D** $\dfrac{35 - 7s}{5}$

   A1.2.2

3. Which numbers in the set $\{0, 13, 15, 21\}$ make the inequality $3 - x > x - 23$ true?

   **A** $\{0, 13, 15, 21\}$
   **B** $\{13, 15\}$
   **C** $\{15, 21\}$
   **D** $\{0\}$

   A1.2.3

4. What is the slope of a line that goes through the points $(13, 17)$ and $(6, 14)$?

   **A** 3

   **B** $-1$

   **C** $\frac{3}{7}$

   **D** $-\frac{3}{2}$

   A1.4.2

5. Solve $3x^2 - 11x + 6 = 0$ by factoring.

   A1.8.2

6. What is the equation of a line passing through the points $(12, 18)$ and $(3, 6)$ in slope intercept form?

   **A** $y = \frac{4}{3}x + 2$

   **B** $3y - 4x = 6$

   **C** $3y - 4x - 6 = 0$

   **D** $x = \frac{3}{4}y - \frac{3}{2}$

   A1.4.3

7. What are the $x$-intercept and $y$-intercept of $y = \frac{7}{5}x - 3$?

   **A** $\left(\frac{7}{5}, 0\right)$ and $(0, 3)$

   **B** $(3, 0)$ and $\left(0, \frac{15}{7}\right)$

   **C** $\left(\frac{7}{5}, 0\right)$ and $(0, -3)$

   **D** $\left(\frac{15}{7}, 0\right)$ and $(0, -3)$

   A1.4.2

8. Solve the equations $2y + 6 = x$ and $y - 3x = -13$ using substitution.

   **A** $(8, 1)$
   **B** $(4, -1)$
   **C** $(3, -4)$
   **D** $(12, 3)$

   A1.5.3

9. Use subtraction to solve the equations: $x + 4y = 10$ and $x + 7y = 16$.

   **A** $(-2, 3)$
   **B** $(2, 2)$
   **C** $(-5, 3)$
   **D** $(-10, 5)$

   A1.5.4

10. What is $\sqrt{x^{13}y^4}$?

   A1.6.3

11. What is the solution to $y = -\frac{2}{3}x + \frac{8}{3}$ and $y = \frac{2}{3}x + \frac{4}{3}$, given the graph below?

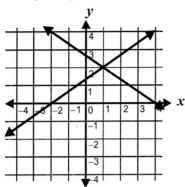

**A** $(4,4)$
**B** $(0,3)$
**C** $(-2,0)$
**D** $(1,2)$

A1.5.1

12. Solve $3x^2 + 5x - 2 = 0$ by factoring.

**A** $x = \frac{1}{3}, 2$

**B** $x = -2, \frac{1}{3}$

**C** $x = -\frac{1}{3}, 2$

**D** $x = -2, -\frac{1}{3}$

A1.8.2

13. Solve $x^2 + 12x - 36 = 9$ by completing the square.

**A** $x = -6, 6$
**B** $x = -15, 3$
**C** $x = -87, 75$
**D** $x = 2, 4$

A1.8.4

14. Simplify $\sqrt{112}$.

**A** $4\sqrt{3}$
**B** $-4\sqrt{7}$
**C** $4\sqrt{7}$
**D** $7\sqrt{4}$

A1.1.2

15. Which of these graphs represents the inequality $3y \leq 2x - 6$?

**A**

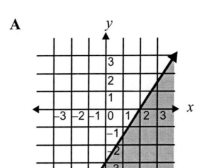

**B**

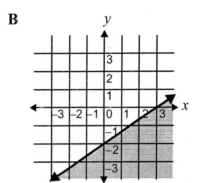

**C**

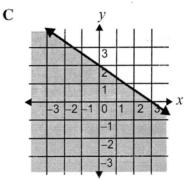

**D**

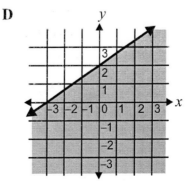

A1.4.6

16. Which of the following inequalities matches the graph below?

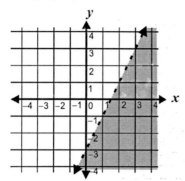

**A** $y > \frac{3}{2}x - 3$

**B** $2x - 3y < 0$

**C** $5y - 10x + 1 < -14$

**D** $5y - 10x + 1 > -14$

17. What is the range of $y = x^2 - 8x + 7$?

**A** $-7 \leq y < \infty$

**B** $-9 \leq y < \infty$

**C** $7 \leq y < \infty$

**D** $9 \leq y < \infty$

18. The food stand in the baseball stadium of Atlanta sells energy drinks and hot dogs. 7 hot dogs and 4 energy drinks costs $15.90. 5 hot dogs and 6 energy drinks cost $15.60. How much does one hot dog cost?

**A** $1.50

**B** $1.35

**C** $0.75

**D** $1.65

19. Factor $6x^2 - 17x + 5$.

**A** $(3x - 1)(2x - 5)$

**B** $(x + 5)(x + 1)$

**C** $2(3x - 5)(2x + 1)$

**D** $6(2x + 1)(x - 3)$

20. Howard tossed a rubber ball into the air. The ball bounced three times and then dropped into a storm drain, Which of the following graphs best models this situation?

**A**

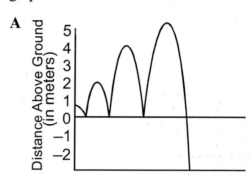

**B**

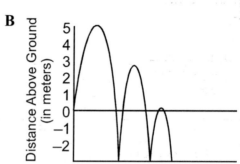

**C**

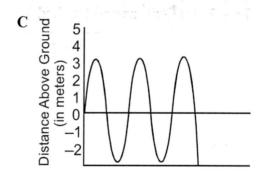

**D**

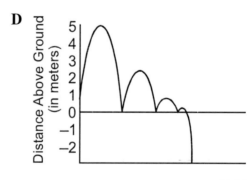

21. Solve the equations $y = 4x$ and $-3x + 2y = 15$ by substitution.

    **A** $(3, 12)$
    **B** $(3, 10)$
    **C** $(5, 15)$
    **D** $(-3, 3)$

A1.5.3

22. Looking at the graph below, what are the zeros of the parabola?

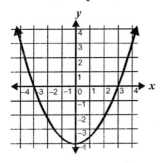

A1.6.8

23. Use a graphing calculator to solve $2x^2 - 7x + 1 = 0$ to the nearest hundredth.

A1.8.9

24. Looking at the graph below, what are the domain and range? Is it a function? Why or why not?

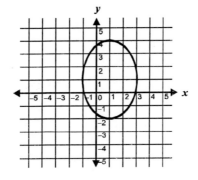

A1.3.4

25. Simplify $\dfrac{x^2 + 6x}{x^2 + 8x + 12}$.

A1.7.1

26. How many meters are in $3,751$ inches? (Note: There are $2.54$ centimeters in 1 inch.)

A1.1.5

# Index

## Available from American Book Company

Passing the ISTEP+ 7 & 8 in Mathematics
Passing the NEW ISTEP+ 9 in Mathematics
Passing the ISTEP+ GQE in English Language Arts
Passing the ISTEP+ GQE in Mathematics

ISTEP+ GQE English Language Arts and Mathematics On-Line Testing

## For Additional Review:

Basics Made Easy: Grammar and Usage Review
Basics Made Easy: Reading Review
Basics Made Easy: Mathematics Review
Basics Made Easy: Writing Review

SAT Test Preparation Guides: Mathematics, Reading, Writing
ACT® Test Preparation Guides: English, Reading, Mathematics, Science

Mathematics Flash Cards
Periodic Table Flash Cards
Science Flash Cards
Projecting Success! Transparency Series: Language Arts
Virtual Math Tutor Software
High School Core Knowledge Software

**AMERICAN BOOK COMPANY**
**PO BOX 2638**
**WOODSTOCK, GEORGIA 30188-1383**
**TOLL FREE: (888) 264-5877**
Visit our web site at www.americanbookcompany.com